Sustainable Value Management for Construction Projects

Ayodeji E. Oke · Clinton O. Aigbavboa

Sustainable Value Management for Construction Projects

Ayodeji E. Oke
Faculty of Engineering and the Built Environment
University of Johannesburg
Johannesburg
South Africa

Clinton O. Aigbavboa
Faculty of Engineering and the Built Environment
University of Johannesburg
Johannesburg
South Africa

ISBN 978-3-319-85333-8 ISBN 978-3-319-54151-8 (eBook)
DOI 10.1007/978-3-319-54151-8

Printed on acid-free paper

This Springer imprint is published by Springer Nature
The registered company is Springer International Publishing AG
The registered company address is: Gewerbestrasse 11, 6330 Cham, Switzerland

To God
and
our family members

Preface

Achieving sustainability has been a global concern for people of various countries, professions and disciplines. In recent years, agencies such as the United Nations (UN) have tailored their goals and objectives to the achievement of sustainability in terms of social, economic and environmental sustainability. For construction, it has become unavoidable to deliver infrastructures that are economically, socially and ecologically sustainable because of various challenges posed by construction activities such as deforestation and air pollution, among others.

There have been many management tools for the effective production of infrastructures in the construction industry and one of them is value management. Value management provides a multidisciplinary framework with the aim of achieving the best function of products and elements at the lowest possible overall cost. It involves the identification of alternatives, and the elimination of unnecessary cost, materials, labour and energy, while striving to provide the best function of the element. Value management has been applied to construction projects in most developed and some developing economies, while it is barely gaining popularity in some other economies. It is therefore necessary to create an awareness of the benefits and understanding of the various obstacles to its implementation among construction industry stakeholders. It is also essential to understand the various drivers that assist in the implementation of the discipline. This book therefore provides readers with the various drivers, barriers and benefits of adopting value management as well as the medium of application of the tool for achieving sustainable construction projects. There are a number of scholarly books in each of these aspects but there is a lack of literature that integrates the link between the two areas as discussed above.

The book adopts divisions into various parts and chapters to highlight various concepts of value management and sustainable construction, which are collectively referred to as sustainable value management for construction projects. Each of the chapters commences with an introduction describing what to expect from the section and concludes with a summary highlighting major issues raised. As this is a research book, references are provided at the end of each chapter for further reading and expansion of knowledge. An index of important and key words is also provided for a quick reference to areas of interest.

The expected readers of this book include built environment scholars; government agencies (public clients) such as parastatals, ministries and other arms of government that are concerned with the provision of infrastructure and other associated developmental projects; corporate agencies involved in planning, executing and managing infrastructures; individual clients who desire value for money for their projects; owners of construction projects; policy makers who are concerned with improving the performance of construction projects; construction professionals tasked with the responsibilities of development and monitoring of construction works; bodies and boards concerned with the monitoring and regulation of the professionals; building contractors in various categories of project execution in building, civil, and industrial engineering areas; and financiers of construction projects, including banks, insurance companies, bond companies, and loan firms, amongst others. It will also be useful for stakeholders in the construction education sector including education boards; principal administrators of education-related institutions; and researchers in the field of architecture, building, construction management, estate management, engineering, land surveying, project management, quantity surveying, urban and regional planning, and other built environment areas.

The book can be adopted as research guide, framework, aid, note or material for topics relating to value management in construction, the concept of construction projects, project performance indices and sustainable developments in the construction industry. We hope that all readers of this book will find it not only stimulating, insightful and impacting but also helpful in advancing their knowledge in the areas of value management and achievement of sustainable developments.

Johannesburg, South Africa

Ayodeji E. Oke
Clinton O. Aigbavboa

Contents

About the Authors

Ayodeji E. Oke is a postdoctoral research fellow in the Department of Construction Management and Quantity Surveying, University of Johannesburg, South Africa and a lecturer in the Department of Quantity Surveying, Federal University of Technology Akure, Nigeria. He obtained his Ph.D. from the Federal University of Technology, Akure and has been teaching several courses in value management, quantity surveying and construction project management for some years. His areas of specialisation are construction value, cost and general sustainability management. He has published scholarly papers relating to these areas in several journals and conference proceedings.

Clinton O. Aigbavboa is Associate Professor in the Department of Construction Management and Quantity Surveying, University of Johannesburg, South Africa. Before entering academia, he was involved as quantity surveyor on several infrastructural projects, both in Nigeria and South Africa. He completed his Ph.D. in Engineering Management and has published several research papers in the area of housing, construction and engineering management, and research methodology for construction students. He has extensive knowledge in practice, research, training and teaching.

Part I
Background Information of the Book

Chapter 1
General Introduction

Abstract This chapter introduces the idea behind the conception of this research book with the emphasis on the explanation of key issues relating to construction project value as well as existing views and perceptions of construction stakeholders towards value management. Various cost-saving practices and techniques available in the construction industry are discussed and distinguished from value management exercise. The objective of the book, which is to promote sustainable infrastructural development in the construction industry, is also explained while detailing the scope and areas of concern that are addressed in each of the book chapters.

Keywords Construction project · Sustainable construction · Sustainable value management · Value · Value management

Introduction

Construction has evolved from ancient times when people built tents using available materials such as leaves, timber, and palm fronds. In the modern era, the quest for complex projects and the changing demands of clients have led to the emergence of various principles and techniques all geared towards meeting the goal of providing infrastructures that are not only aesthetically relevant, but that meet the existing as well as the future needs of society. At the early stage of modern construction, especially after World War II, individuals referred to as contractors came into existence and they were usually invited to bid for reconstruction and new works. This gave rise to various challenges, including the management of building works and stakeholders involved in the construction process necessitating the introduction and adoption of various management mechanisms, such as the use of an estimator to assure fair competition among contractors. Over time the contractors adopted their estimators for better pricing to increase their chance of winning contracts.

Traditional measures of project success are related to time, cost, and quality as well as other techniques introduced at the earliest stage of construction projects to

A.E. Oke and C.O. Aigbavboa, *Sustainable Value Management for Construction Projects*, DOI 10.1007/978-3-319-54151-8_1

ensure that the proper measurement of these measures are related to the three areas. These earlier techniques included feasibility and viability studies, cost control, and quality management, among others. There is increasing complexity of the construction industry as a result of several factors. Stakeholders including professionals and non-professionals, are now involved in the conception, construction and post-construction phases of projects while the discovery and proliferation of new and innovative construction materials is another contributing factor. Innovation in the making and usage of construction-related machines and equipment has made possible that which initially seemed to be impossible. Likewise, the construction process is becoming more competitive as a result of new contractors' innovative practices in their quest to gain popularity and competitive advantages in construction business. Moreover, the changing business environment and the way of life of various stakeholders have evolved over time, and construction of infrastructural projects are expected to align with reality. In view of these factors, amongst others, new techniques of managing construction projects have emerged and been introduced. These include facility management (FM), value management (VM), lean management (LM), building information modelling (BIM), and other forms of construction management tools. It should be noted that these new and emerging techniques are directly linked to the changing measures of project performance.

Currently, the performance of construction projects has gradually shifted from cost, time, and quality to other indices such as clients' satisfaction, stakeholders' satisfaction and various sustainability measures. These include green buildings and construction, intelligent infrastructures, as well as smart cities and housing as explained in Chap. 5 of this book. In order to meet these targets, there has been anxiety in some quarters regarding the ability of the construction industry to respond and proffer better solution in this regard. However, there are some existing construction management techniques that have not been fully explored and utilized in the industry. One of the techniques is the value management concept. This book is geared at exploring the adoption and usage of value management techniques for the actualisation of sustainable infrastructural development in the construction industry.

The book is divided into five parts and twelve chapters for guidance and ease of use. The first part consists of a chapter detailing the background information for the book, while the issues of value management in construction are discussed in the second part. The second part of the book describes value management in construction. This is further sub-divided into two chapters, namely the history of value management and the concept of value management as a discipline. Part three of the book explores information relating to sustainable construction and three chapters have been allocated to deal with this concept. The first chapter addresses general issues relating to construction projects, followed by a chapter describing various measures of project success and the last chapter entails a discussion of sustainability in construction. The fourth part of the book consist of three chapters that relate specifically to value management and sustainable construction. The chapters in this section discuss value management as a construction management tool; value management as a construction sustainability tool; as well as a detailed discussion of the stakeholders in sustainable value management in construction. After various explanations of the two main areas and their links, the concluding part relates to the enhancement of value management in the construction industry. Likewise, this section of the book consists of three chapters that address the challenges, drivers, and prospects of adopting sustainable value management in the construction industry.

The first chapter of the book introduces the concept of value management, covering aspects such as the meaning of value in construction, perceptions and motives of value management, cost saving practices in the construction industry, as well as the general objective of the book. In Chap. 2 of the book, the history of value management is discussed with the emphasis on the origin and aim of the practice; the distinction between value management, value engineering, and value analysis is also discussed. This section further examines the current state of value management adoption in the construction industry; areas of application of value management in construction; as well as risks and cost of applying value management for construction works. Chapter 3 of the book explains the concept of value management as a discipline by extensively discussing the procedure of value management; the facilitation process; function analysis as it relates to value management; life cycle costing; function versus value of construction works; value, cost, worth and price; and the concept of unnecessary cost in construction projects.

In discussing the concept of sustainable construction, a detailed content of construction projects is explained in Chap. 4 with the emphasis on construction as a process, the participants and stakeholders in the execution of construction projects, as well as the overall objective of the construction industry. Further to this, Chap. 5 details various indices of measuring project performance. The variables discussed include cost, time, quality, energy efficiency, stakeholders' satisfaction, and other emerging metrics. One of the current measures of project performance relates to sustainability and Chap. 6 of the book extensively addresses the idea of sustainability in construction. The chapter discusses the history and origin of sustainability, construction sustainability, various elements of sustainable developments, the current level of adoption of sustainable construction, as well as the drivers, barriers and benefits of adopting sustainable construction.

In exploring value management as a sustainable tool, Chap. 7 of the book entails discussions relating to the adoption of value management as a construction management tool. Several areas of construction management practice are explored with the emphasis on how value management can be adopted. Aspects addressed in this chapter include risk management, project management, lean management, asset management, total quality management, as well as knowledge management. Chapter 8 is the core of the book, it is designed to link the principle of value management to the three-pod of sustainable construction, which are social, economic and environmental aspects. The practice of value management is explored as a financial, social, and environment sustainability tool. To further discuss the link, Chap. 9 highlights and explains various stakeholders to value management and sustainable development in the construction industry. Some of the stakeholders discussed include the clients, consultants, owners, sponsors or financiers, and statutory bodies, amongst others.

To enhance and increase the practice of value management for sustainable construction, Chap. 10 of the book is designed to address the challenges and barriers to the practice of sustainable value management. These include the timing of the process; approaches to the value management workshops; team members' composition; wrong views and perceptions of the discipline; project procurement, tendering and contractual methods; as well as other indirect factors that affect value management in practice. Chapter 11 focuses on various drivers to the adoption of value management in the construction industry. Issues discussed include stakeholders' readiness and participation; existing and emerging value management policy, guidelines and regulations; the level of awareness and knowledge of value management; training and education; the electronic approach to value management, and modern approaches to value management workshops, while other drivers is used to explain additional means of encouraging the practice of value management. The last chapter of the book discusses the prospects of adopting value management for construction works. The benefits explained include innovations, elimination of unnecessary costs, whole life cycle consideration, early problem identification, adoption of new materials and technologies, value for money, and other indirect benefits.

Value in Construction

In the conceptualisation of the theory of economy, one of the areas of contention is the distinction between value, cost, worth, and price in relation to commodity, services, goods, and entity. However, depending on the perspective of individuals, several authors have been able to distinguish and differentiate these as explained in Chap. 3 of this book. In the construction industry, the common services are related to professional responsibilities while the goods being procured are the construction projects which are executed or constructed through established process, standards, and principles. On issues relating to value in construction, authors such as Kashiwagi and Savicky (2003), Jaakson (2010), Salvatierra-Garrido and Pasquire (2011), and Jollands et al. (2015) have discussed concepts of value as they relate to construction projects from the perspectives of various stakeholders. The concept of value as observed by The Institute of Value Management (2015) is concerned with the connection between various needs of the project and the resources that are appropriate in addressing and meeting the needs.

The concept of value has gained popularity and momentum in the construction industry and various research studies are being conducted in the area (Perera et al. 2011). Salvatierra-Garrido and Pasquire (2011) note that value in construction is related to such indices as function, cost, and quality with a further explanation that there is currently no universal definition of the term in the construction industry. In the view of Kashiwagi and Savicky (2003), there is a lack of understanding of the concept of value among construction projects owners and this has resulted in their unwillingness to pay more to attain it. Explaining the concept of best value from a procurement perspective, the concept was linked to a reduction of first or transaction cost of delivering construction projects. One of the major issues relating to value is the ability to differentiate it from cost. Most construction stakeholders, especially owners and financiers, are still cost conscious and achieving value for construction projects is usually associated with cost. However, the issue of value goes far beyond cost: in some cases, it can lead to more or less costs against those which were initially planned: it is more closely related to function and the whole-life cycle cost of the project.

The concept of value is subjective in nature depending on the view, desire, objective, or target goal of particpants. In the construction industry, the participants are various direct and indirect stakeholders who are shouldered with the responsibilities of ensuring that the project performs to expectation. This implies that value attached to construction projects depends largely on the expected outcome of the participants. Value in construction projects is therefore related to all performance indices, including the traditional and emerging indices. These include cost, time, quality, satisfaction, and sustainability, among others. Value is therefore not about one of the indices, but the entire measures of project performance that result from the purpose of the project as agreed by participants at the conception stage or the established standards by bodies and agencies shouldered with the responsibilities of managing and regulating construction projects within a specified domain.

An important aspect of achieving value in construction is that participants should understand the indices of its measurement as this affects the acceptance level. For instance, in using the teamwork principle, one of the goals of value management practice is to first sensitize members of the workshop on issues relating to projects under consideration and the purpose of the practice. This thus helps the team members to have the same focus against what is perceived by individuals by virtue of their training, specialty, experience, and interest.

Perceptions of Value Management

Value management principle was first introduced in the manufacturing industry but it has gained popularity in other sectors, including the construction industry, especially in developed and some developing countries. Owing to the lack of knowledge and resistance to change, value management has been construed to have different meanings. Some perceive it to be an extension of feasibility and viability studies while others acknowledge it as another cost saving or cost control exercise. Some researchers and stakeholders in the construction industry have discarded the practice, pointing to its similarities with existing principles such as project management, risk management, cost management, and the like. However, in countries where there is a proper understanding of the practice, value management has contributed significantly to performance and the better delivery of construction projects. In this book, various issues surrounding value management are explained and various wrong perceptions that have hindered the practice of value management in some countries are highlighted in Chap. 10.

Owing to various views and perceptions of value management, there are numerous recognised but contentious names for the disciplines (Palmer et al. 1996; Male 2002; Kelly et al. 2014; Society of American Value Engineers 2015; The Institute of Value Management 2015). They include the following:

- Value management;
- Value engineering;
- Value analysis;
- Value assurance;
- Value planning;
- Value improvement;
- Value methodology; and
- Value control.

However, judging by studies in the area by previous researchers, the term 'value management' has been adopted for this book. The reasons for the choice as well as an explanation of the other terms are given in detail in Chap. 2.

Project Cost Saving Exercise

One of the earlier practices in the construction industry is the saving of construction cost for clients, owners, and sponsors of construction projects. It was believed that clients would be satisfied and appreciate the consultants and other stakeholders of their projects if what was expended was less than what had been budgeted for or proposed at the inception of the project. Most of the earlier construction management tools and approaches were tailored to achieve this as well as the other two measures of project success, namely time and quality. These three indices of project performance are still relevant and will continue to be. Recent events in the construction industry indicated that some clients are less interested in these traditional indices of project performance, with the emphasis on other measures. Chapter 5 of this book examines various measures of projects success including those which are old, current and emerging.

Value management can be a cost-saving exercise as observed from construction projects where the principle has been adopted, but a careful observation of previous projects where the principle was adopted has shown that the objective of the application of the principle is beyond cost saving of construction works. In fact, the principle behind its introduction is more of the general management of project success than that of focusing on cost. This is one of the wrong perceptions of the practice, this is further explained in Chap. 10 as one of the barriers to the adoption of the practice of value management in the construction industry.

Objective of the Book

There is plenty of existing material on value management, value management in construction, sustainable development, sustainability in construction, and other related areas (Norton and McElligott 1995; The College of Estate Management 1995; Palmer et al. 1996; Abidin and Pasquire 2005; Khatib 2009; Kibert 2012; Kelly et al. 2014; Aghimien and Oke 2015; AlSanad 2015; Society of American Value Engineers 2015; The Institute of Value Management 2015; Karunasena et al. 2016; Kibwami and Tutesigensi 2016; among others). These materials include textbooks, research books, book chapters, journal articles, and conference papers that provide the necessary and detailed information regarding each of the identified areas. This book is not designed to repeat the same but to extend and expand existing knowledge on the areas.

This research book on the application of value management as a construction sustainability management tools does not only describe practical issues concerning the two major areas of the 21st century construction goals, but also suggests ways through which value management can bring about sustainable construction. The book explains sustainable value management, not just as a cost saving tool as has been perceived by many researchers, but as an effective tool for construction project performance that will not only ensure that construction projects are viable but will also contribute positively to the environment and be beneficial to society at large. The book also provides guidance on value management and construction sustainability for construction professionals, employers of labour, researchers, and students alike. This book will enhance a better understanding of the concept of value management, the basics of sustainable construction as well as the means of using the principle of sustainable value management to achieve successful construction projects that will be financially viable, socially beneficial, and environmentally relevant and sensitive for the society.

Summary

This chapter introduced the idea behind the conception of this research book with the emphasis on the explanation of key issues relating to construction project value as well as existing views and perceptions of construction stakeholders of value management. Various cost-saving practices and techniques available in the construction industry were discussed and distinguished from the exercise of value management. The objective of the book, which is to promote the application of value management for sustainable construction, was also explained while detailing the scope and areas of concern that are addressed in each of the book chapters. The concept of value management is further explained in chapter two.

References

Abidin, N. Z., & Pasquire, C. L. (2005). Delivering sustainability through value management: Concept and performance overview. *Engineering, Construction and Architectural Management, 2*(2), 168–180.

Aghimien, D. O., & Oke, A. E. (2015). Application of value management to selected construction projects in Nigeria. *Developing Country Studies, 5*(17), 8–14.

AlSanad, S. (2015). Awareness, drivers, actions, and barriers of sustainable construction in Kuwait. *Procedia Engineering, 118*, 969–983.

Jaakson, K. (2010). Management by values: Are some values better than others? *Journal of Management Development, 29*(9), 795–806.

Jollands, S., Akroyd, C., & Sawabe, N. (2015). Core values as a management control in the construction of 'sustainable development'. *Qualitative Research in Accounting & Management, 12*(2), 127–152.

Karunasena, G., Rathnayake, R. M., & Senarathne, D. (2016). Integrating sustainability concepts and value planning for sustainable construction. *Built Environment Project and Asset Management, 6*(2), 125–138.

Kashiwagi, D., & Savicky, J. (2003). The cost of 'best value' construction. *Journal of Facilities Management, 2*(3), 285–297.

Kelly, J., Male, S., & Graham, D. (2014). *Value management of construction projects*. Oxford: Wiley.

Khatib, J. (2009). *Sustainability of construction materials*. London: Elsevier.

Kibert, C. J. (2012). *Sustainable construction: Green building design and delivery*. Washington: John Wiley & Sons.

Kibwami, N., & Tutesigensi, A. (2016). Enhancing sustainable construction in the building sector in Uganda. *Habitat International, 57*(2016), 64–73.

Male, S. (2002). A re-appraisal of value methodologies in construction. *Construction Management and Economics, 11*(2002), 57–75.

Norton, B. R., & McElligott, W. C. (1995). *Sustainability of construction materials*. London: Macmillan Building and Surveying series.

Palmer, A., Kelly, J. & Male, S. (1996). Holistic appraisal of value engineering in construction in United States. *Construction Engineering and Management*, 324–326.

Perera, S., Davis, S., & Marosszeky, M. (2011). Head contractor role in construction value-based management: Australian building industry experience. *Journal of Financial Management of Property and Construction, 16*(1), 31–41.

Salvatierra-Garrido, J., & Pasquire, C. (2011). Value theory in lean construction. *Journal of Financial Management of Property and Construction, 16*(1), 8–18.

Society of American Value Engineers. (2015). *What is value engineering?*. NJ: Mount Royal.

The College of Estate Management. (1995). *Value engineering*. UK: Reading.

The Institute of Value Management. (2015). *What is value management?*. UK: Ledbury.

Part II
Value Management for Construction

Chapter 2
The Concept of Value Management

Abstract The concept of value management was introduced to compare alternative materials in order to arrive at the one that provides the best function at the lowest possible overall cost. This chapter discusses the initial introduction of the concept to the manufacturing industry but indicates that it has since gained popularity in other sectors, including the construction industry. Owing to the varying knowledge and different levels of perceptions of the essence and principles of operation of the practice, various terms were attributed to the discipline but the most common and all-encompassing one is that of 'value management'. As beneficial as the practice has been in countries and projects where it has been adopted, there are some risks that required attention to achieve the purpose for which it was conceived. The risks, as well as direct and indirect costs of conducting the exercise, are also identified and discussed in this chapter.

Keywords Construction project value · Value analysis · Value engineering · Value management · Value management cost · Value management risk

Introduction

Every concept, procedure and practice must have been conceived, actualised, and practised at one time or the other by an individual or a group of people in an informal or formal establishment, organisation or industry. However, in some cases, there could have been an improvement in way the concept is practised due to innovation and the changing nature of people and concern for the environment. It is possible that the original name of the concept may have been changed, transformed, or modified for obvious reasons. In this chapter, the history of value management—in general and in relation to the construction industry—is discussed with the emphasis on the original goal, various names that have been and currently are being used for the concept, as well as the current level of adoption of the practice in the construction industry. Various areas of application of the discipline, risks associated with the practice, and the cost of organising value management are also discussed.

A.E. Oke and C.O. Aigbavboa, *Sustainable Value Management for Construction Projects*, DOI 10.1007/978-3-319-54151-8_2

History of Value Management

Various authors have contributed their opinions as to the original source of value management and there seems to be consensus in their views (The College of Estate Management 1995; Green and Moss 1998; Palmer et al. 1996; Finnigan 2001; Male 2002; Shen and Liu 2004; Liu and Shen 2005; De Leeuw 2006; Kelly and Male 2006; Short et al. 2008; Oke and Ogunsemi 2011; Perera et al. 2011; Shen and Yu 2012; Leung and Yu 2014; Jay and Bowen 2015; Society of American Value Engineers 2015; The Institute of Value Management 2015; Karunasena et al. 2016). Value management (VM) was first introduced in the United States of America around the 1940s and various authors have attributed the credit to an employee (Lawrence Miles) of the General Electric Company who was at the time working as a purchase engineer. This implies that the practice was first introduced to the manufacturing industry during World War II owing to the scarcity of basic resources for the manufacturing of goods for human consumption. The shortage was so acute that it was becoming difficult for the industry to meet the requirements of society. As a result, Lawrence Miles came up with a concept of finding alternative materials, components, or resources that could perform the function of unavailable ones at the least possible cost.

The first known research material in the subject of value management is that of the work of Lawarance D Miles (1972). However, several adjustments and modifications have been made to the initial principles of value management. In the case of the construction industry, these include the following:

- Improvement in the definition of value to include such elements as customer satisfaction, user satisfaction and currently sustainability issues;
- Change in the adoption of value engineering and value management in place of value analysis;
- Application in several sectors, numerous ways and for various purposes while still retaining the principle of best function at the least cost;
- Advent of systematisation methods for better analysis of alternative ideas such as the function analysis system technique (FAST);
- Improvement in decision-making techniques, from brainstorming to such principles as simple multi-attribute rating technique (SMART);
- Adoption of whole-life costing in the evaluation of a project beyond the completion of the projects, but throughout its lifespan;
- Introduction of pre-study and post-study phases at the beginning and ending of value management study respectively;
- Application and adoption of other management principles such as lean construction, total quality management; and
- Modification of initial 40-h workshop to other new techniques.

Several countries have also adopted value management in various sectors of their economy. For instance, the practice was first introduced to manufacturing companies owned by the Chinese states in 1978; Australia adopted it through the activities of some multinational companies in 1960s; it was pioneered in Hong Kong in 1988, while it was introduced in Nigeria in the 1990s through workshops, seminars, and conferences organised by stakeholders in manufacturing, production, and the construction industry (Liu and Shen 2005; Shen and Yu 2012; Oke and Ogunsemi 2013). The following are some of the value organisations tasked with the responsibilities of regulating value and value management practices in some countries and their year of establishment:

- United States of America—Society of American Value Engineers (SAVE, 1959)
- Japan—Society of Japanese Value Engineering (SJVE, 1965)
- United Kingdom—The Institute of Value Management (IVM, 1965)
- Germany—Vereine Deutscher Ingenieure/Gesellschaft Systementwicklung Wirtschaft (VDI-GSP, 1967)
- India—Indian Value Engineering Society (INVEST, 1977)
- Taiwan—Value Management Institute of Taiwan (VMIT, 1977)
- South Korea—Society of Korean Value Engineers (SKVE, 1983)
- Brazil—Associacao Brasileira de Engenharia e Analise do Valor (ABEAV, 1984)
- Canada—Canadian Society of Value Analysis (CSVA, 1984)
- The Netherlands—Dutch Association of Cost Engineers, Special Interest Group Value Management (DACE, 1984)
- Hungary—Society Hungarian Value Analysis (SHVA, 1990)
- Saudi Arabia—Arabian Gulf Chapter (AGC, 1990)
- Spain—Associacio Catalana d'Analisi del Valor (ACAV, 1990)
- Australia—Institute of Value Management (IVM, 1991)
- France—Association Francaise pour l'Analyse de la Valuer (AFAV, 1993)
- Hong Kong—Hong Kong Institute of Value Management (HKIVM, 1995)
- Portugal—Associacao Portuguesa para a Analise do Valor (APAV, 1998)
- Malaysia—Malaysian Institute for Value Management (MIVM, 1999)
- China—Value Engineering Society of Beijing (VESB, 2001)

Aim and Definition of Value Management

According to Miles (1972), value management was introduced to examine and analyse alternative materials for the purpose of selecting the one that provided same, better, or best function at the least cost. Value management was conceived and practised at the early stage of project conceptualisation as a result of the need for innovation, novelty and advancement of existing practice. There are possibly different views of values from various construction participants, but the aim of value management practice is to unify these differences in order to achieve the project's stated goals using minimum resources (The Institute of Value Management 2015). VM entails using every possible resource and opportunity to improve the value of a component (system, material, element or resources).

The objective of VM therefore is to provide the best function at the lowest possible overall cost. As simple as this may be, some important variables are considered to be of importance to the practice of value management (Miles 1972; Palmer et al. 1996; Shen and Liu 2004; Short et al. 2008; Hewage et al. 2011; Perera et al. 2011; Shen and Yu 2012; Oke and Ogunsemi 2011; Oke and Ogunsemi 2013; Kelly et al. 2014; Oke et al. 2015; Yekini et al. 2015). These include the following:

- Management process;
- Systematic approach;
- Teamwork and multi-disciplinary principle;
- Analysis of function technique;
- Best value concept;
- Stages of project or product;
- Whole-life cost concept;
- Value for money principle; and
- Issues relating to investment return.

Value management is a management process that involves the control, monitoring and managing of project team members, redesigning of spaces and components, appropriate selection of materials, as well as the optimisation of the process of producing a product in order to meet the stated project goals. It is a holistic process of managing all forms of resources and this differentiates the practice from other cost-cutting or cost-saving exercises. The second point is that value management adopts a systematic approach that is logical, methodological and organised so that members of the team can easily participate and the approach can be adapted for subsequent exercises. There are various approaches to conducting a value management workshop, but regardless of the approach, an agreed laid-down principle must be followed in planning, organisation, conducting the actual workshop, and reporting the findings or recommendations to clients or stakeholders or commissioned agencies through an appropriate feedback mechanism.

Every technique or management activity has its way of operation and in most cases, a particular set of people are involved because of their profession experience, job specification, or employment status, among others. However, value management is a multi-disciplinary technique where people and individuals from various backgrounds and of diverse dispositions are selected and assembled in a team to brainstorm and agree on the optimisation of function and cost. Although there are various suggestions regarding the set of people that should form a value management team, it is worth noting that a successful value management exercise involves both professionals and non-professionals in construction and other concerned fields. In construction related activities, for instance, it is necessary to engage stakeholders with knowledge of construction who are mainly participants and stakeholders in the industry. These stakeholders include construction professionals, contractors, value management expert among others. The stakeholders should also include some members of the original design team for ease of application of the recommendations emanating from the value management exercise to the main work.

Further to the engagement of these set of stakeholders, modern value management principles also advocate for the involvement of non-professionals that are able to bring a 'non-professional perspective' to the project. Moreover, there are projects stakeholders who are directly or indirectly impacted by the execution of projects. For instance, an accountant who is a non-construction professional may be involved as a member of the construction value management team and it is unlikely that some of the suggestions and ideas coming from this source may be more beneficial to the project than those of the professional members. In a shopping mall project, for instance, it is not out of order to involve the users, that is, individuals who will eventually rent and use the facilities, in the value management team. It may also not be surprising that some of the elements that are part of the original design may be of no value to them, that is, may perform no function, which is the principle on which value management is based.

An important element of value management is the principle of functional analysis. This stems from the fact that the function of a product or element may differ, depending on the point of view of the individuals concerned and the purpose of the project under consideration. For instance, a structural column at the entrance of a building may be viewed from different angles by various construction professionals. To an architect, it may have been introduced for aesthetics, whilst an engineer will be concerned with the structural aspect of the column, and the quantity surveyor (cost estimator) will likely focus on the cost implication of such a column. To guide against misconceptions and individual views, it is fundamental for a value management team to highlight all the elements or components of a project or process and their functions. These functions can be identified in term of primary (main) and secondary ones which are related to quality, reliability, performance, satisfaction, and the like This thus helps members of the value management team to focus on a particular and unified direction while discussing a specific element or components of a project.

The best value concept is linked to functional analysis, whole-life cost, return on investment, and stages of the project under consideration. Value has been explained in chapter one of this book under value in construction. It is also discussed in relation to cost, price, and worth in Chap. 3. The value of a product or element is related to the function and cost. This implies that for every product or element, there are some unnecessary costs as a result of unnecessary materials, activities or processes that add little or nothing to its function. Eliminating these costs will help in providing best value and subsequent value for money for clients and stakeholders. The best way to determine unnecessary cost in a product or element is by identifying alternative materials that provide the same, or nearly the same, function or serve almost a similar purpose as the item under consideration. Comparing the elements or products from the same view of the function they are meant to provide based on their life cycle cost will help in selecting the best alternative material that will give the best value.

In the evaluation of alternatives, one of the principles of value management is the consideration of different stages of a project and how the alternatives materials will fit. For instance, in construction, the stages of a construction project are from conception to inception, planning, actual construction, completion, usage, conversion, and demolition. The best alternative materials should not only be the cheapest but the duration or stages of the main project should be a major deciding factor. A temporary project that is built to last for just two years will require materials for the same lifespan. This leads to the principle of re-using materials which is more related to whole-life costing.

The principle of considering the stages of a project is that the best alternative material should be adopted, provided it will last the duration of the main project. Alternatively, cheaper material could be considered even if it has to be replaced before the end or demolition of the main project. Whole-life costing is related to all associated costs of a product or element, including initial, running, and demolition or re-use costs. The term has evolved over the years, from 'life costing' to 'life cycle costing', 'whole-life costing' and to the currently favoured, 'whole-life cycle costing'. The principle is that it will be unfair to judge products or elements only by their initial or running cost but the total costs that will be required for the entire life of the project should be considered. This is where the term 'overall cycle cost' associated with value management is derived from.

Every project is a form of investment with an expected direct or indirect return. The return can be in the form of cost-related gains, services for people, and meeting the need of individuals, which is related to the principle of satisfaction. Value management takes into consideration value for money and return on investment of clients, owners, financiers or sponsors of the project using various investment-related approaches. This aspect also considers the whole life of the project, component, or product in judging and selecting an appropriate investment appraisal technique. Ranging from the simple concepts such as pay-back period, internal rate of return, cost-benefit analysis, to the recent computer-based algorithm methods, the essence is to use mathematical principles to compare alternative materials with the notion of selecting the one with best value that provides the best return on investment.

Judging from the key principles of value management, this book therefore defines the concept of value management as a systematic and methodological project management process that adopts a diverse and multi-disciplinary approach of analysing the function of elements or products, using the whole-life cycle principle, through the stages of a project for the purpose of achieving the best function of the whole project at the lowest possible cost, and thereby enhancing best value and better return on investment. The earlier explanations of key areas of value management suggest that, for the process to be effective and successful, it requires a basic understanding of the principles by team members, especially the facilitator.

Value Management, Engineering and Analysis

As discussed in chapter one under 'perception of value management', this section explores the concept of the practice with the emphasis on the origin and meaning of various terms that have been used to describe VM. Such terms include value engineering, value planning, value analysis, value control, value methodology, value improvement, and value assurance. Perera et al. (2011) note that value management originated as value analysis in the United States of America. However, over the years, authors from the region have adopted the use of value engineering to describe the term, with the concept adopting the same principles of value management explained earlier under the 'aim and definition of value management' section of this chapter. The Society of American Value Engineers (SAVE) has adopted the term 'value engineering' as observed from their name and public reports published by the body over time. This implies that the term has metamorphosed from 'value analysis' initially proposed by Lawrence Miles during the World War II to 'value engineering' in the United States of America.

Published materials from the United Kingdom (UK) have adopted 'value management' to describe the process. A similar body to SAVE, that is, The Institute of Value Management (IVM), is also tasked with the responsibility of controlling the practice and process of the discipline in the UK. Authors affiliated with the USA, especially those from countries that adopt American English, prefer the term 'value engineering' while those from regions and countries with links to the UK and adopt British English usually use the term 'value management'. Only very few recent sources adopt any of the remaining described affiliated terms while most authors described them to be part of the main process of value management, or value engineering, as the case may be.

Value Methodology

Value methodology refers to the process, principles and techniques adopted in the conduct and practice of value management and it includes those practised at value planning, engineering, and analysis phases.

Value Planning

Value planning is an aspect of value management that is associated with achieving project value during the planning stages of a project. For instance, in construction this is associated with value at the early stage, namely, conception, inception, feasibility, viability, and other planning-related activities of the project. Value planning is a sub-set of value control and they are both derived from the principle of cost planning and cost control, which are common terms for management of developmental projects.

Value Control

Value control in respect to value management is concerned with managing value throughout all stages of a project where cost control is practised. This indicates a direct link to cost control which is not the same as value management, justifying the reason why the term is not common among value management experts, analysts, or researchers.

Value Analysis

Value analysis is associated with the post-construction or completion phase, indicating that the practice is related to the value of completed project. This is inclusive of the use and re-use stages of a project. Management is a general term inclusive of improvement and assurance, implying that 'value improvement' and 'value assurance' are synonymous with value management.

Value Engineering

The closest term to 'value management' is 'value engineering' which is described as the study of value at the design, construction, and engineering stage of a project (Finnigan 2001). Moreover, De Leeuw (2006) concludes that judging from the principle surrounding the concept and conduct of the practice, it is more related to 'value' and 'management' than that of 'value' and 'engineering'.

Recent scholarly works from some authors reveal the use of the above identified terms, e.g. 'value planning' by Karunasena et al. (2016) and 'value methodology' by Leung and Yu (2014). However, most authors agree that value management encompasses other terms in that each of the remaining terms are related to value management at specific stage(s) of a project (Perera et al. 2011; Oke and Ogunsemi 2013; Karunasena et al. 2016). In view of this, value management explains all the concepts, principles, processes, and participants required for each stage of the discipline in a project. This is necessary for the successful implementation of the discipline and also underscores the reason for its adoption in this research book.

Value Management and Construction Industry

Value management was first introduced to the manufacturing industry but it has gained wide popularity in other sectors of the economy, including the construction industry. It was introduced to construction in the USA and UK in the 1960s and 1980s respectively (De Leeuw 2006; Kelly and Male 2006; Perera et al. 2011; Kelly et al. 2014). Before the introduction of value management in the UK, the existing practice of cost planning and control and its importance and popularity among stakeholders made the acceptance of value management difficult. For this reason the practice of value management did not enjoy the attention and popularity in the UK as it did in the USA.

As a result of the benefits associated with value management in the USA and UK, the practice has spread to other continents, regions, and countries of the world. While it has gained popularity in most construction industries of developed countries, it is worth noting that it has not been fully embraced in some developed and most developing economies (Shen and Liu 2004; Liu and Shen 2005; Oke and Ogunsemi 2011; Jay and Bowen 2015). The possible barriers, drivers, and benefits (or prospects) of the practice for its acceptance and rejection in some national states are discussed in part five of this book, encompassing Chaps. 10, 11, and 12 respectively.

Clients, professionals, contractors and other stakeholders in the construction industry have benefitted considerably from the application of value management where it has been accepted and adopted. In the construction industry, value management can be described as an orderly, organized, and systematic construction project management process that adopts a multi-disciplinary and diverse approach, geared towards the achievement of best functions of elements, materials, and other construction resources at the least possible overall cost, with the aim of maximizing return on investment and realizing best value for money for construction clients.

Risks of Value Management in Construction

There are risks inherent in every activity, including the ones that are conceived for the purpose of managing or controlling people, activities or processes. Management principles are combination of various tools, techniques, procedures and methods for the purpose of planning, controlling and regulating people, events, procedure or practice. It is expected that some risks are inherent in the adopted tools which form part of the risks of the management principles. As beneficial as value management has proven to be, there are some risks that need to be monitored, evaluated, and responded to for the process to be smooth, worthwhile and successfully applied in the construction industry.

Sufficient Time for the Study

One of the major risks to the adoption and acceptance of value management in the initial period is the concern over the period of time it takes to complete an exercise. The earlier approach, and still the most common, is the 40-h workshop which is about a working week (using normal eight working hours in a day). It has been argued that most intending value management participants find it difficult to be away from their day-to-day activities for a whole week and as a result, several other approaches have been introduced such as a three-day workshop. However, for a value management exercise to be meaningful and productive, adequate time should be devoted to complete each of the recognised stages.

Completeness of Cost Information

Another risk associated with value management practice is the lack of full and necessary information of all the costs associated with an element or component. The use of some investment appraisal techniques with the emphasis on some economic factors has helped to mitigate this risk but the fundamental principle of all project planning and evaluation techniques, including various investment appraisal and life-cycle costing, is based on the probability of occurrence. Issues of cost become more problematic in economies where it is difficult to predict inflation and other economic indices that affect the cost and price of materials and resources. As a result of these challenges, it is possible to make incorrect assumptions regarding alternatives which can mar the success of the value management exercise.

Related Functions of Components or Elements

Items or elements with related function are usually difficult to quantify. This is for two reasons: there are components or elements of building that serve more than one unique function and it is sometimes difficult to determine their primary role for specific projects as it may vary from one project to another. A very good example is a window unit which serves the purpose of ventilation, aeration, and a source of natural lighting. The secondary functions might include sound insulation, or water insulation, depending on the position. The second aspect relates to two or more elements performing the same function: it is possible they all perform the function as a primary role or it may be a secondary function to one or some of the elements. Common in this category are wall, doors, and windows. Regardless of where they are located, externally or internally, the wall has its basic function but windows and doors are expected to perform a wall function in addition to their original function. The two scenarios sometimes make it difficult for a value management team to apportion the right function to specific elements, which is a potential risks that can render the whole process meaningless and unsuccessful if not properly examined.

Uniqueness of Construction Projects

No two construction projects are the same, even if they possess the same characteristics in terms of similarity, same client, same contractor, located in the same area, and procured through the same means. The distinctiveness of a project affects the function of some elements and components of the project. For example, a column can be introduced as an aesthetic feature for a project, but may also serve as one of the main structural members of another. This implies that the value management team should be able to identify the functions of various elements bearing in mind the uniqueness of the project in consideration.

Quality of End Product

Predicting the end-product quality of an element or component is one of the tasks that must be given the necessary attention in determining a product that will perform the best function at the least possible cost. This is better achieved through examination of current or past situations where the product has been used. However, there are three major issues with this method. The first concern is the availability of historical data and the required information on the product; another is the analysis of the difference in exposure to weather and other conditions between the present or past and the current situation while the last issue relates to examining the conditions of the use of the products.

Selection of Team Members

A major attribute of value management is that it is multi-disciplinary in that members of the team are varied and unlikely to be drawn from the same area of practice or sector of the economy. However, the risk associated with this attribute is the selection of the wrong people who may contribute nothing or negatively to the success of the practice. It is better if the team members accept the practice and have previous experience of value management workshop but the latter criterion should not be mandatory, depending on the nature of project and the set of people to be involved. Some project stakeholders from the design team are expected to be members of the team: as much as it may be difficult to choose from these people, it is more difficult to select other members who were not part of the original design team. As explained under the 'aim and definition of value management' section of this chapter, the other members may be professionals and non-professionals. The onus of the selection therefore lies with the facilitator and this must be done carefully to guide against a choice of members who will add no value to the exercise.

Representation of Original Design Team

For an effective value management exercise, it will be better if some of the existing design team members of the project under consideration are included as members of the team. This can contribute positively to the exercise through the generation of novel idea and ease with which the value management outcomes can be adopted for the project. However, it becomes a risk if the existing team members are unwilling to agree and accept changes to the project that arise from the exercise. Another issue is the establishment of criteria for the selection of members of the original team since not all can be invited to be part of a value management team. The number will be too large, which is one of the barriers to an effective team. The cost of paying for their service is another factor to be considered. It is therefore necessary to ensure that not only 'important' or 'influential' members of the design team are selected but that those who have the time and ability to contribute positively to the success of the exercise are considered as members.

Choice of Facilitator

Value management facilitators are value management team leaders who possess the necessary leadership, team control, and other management skills, competence, and attributes necessary to conduct a value management exercise. One of the key characteristics expected of a facilitator is the knowledge, understanding, and experience of a value management workshop. Depending on the value management methodology adopted, such a person should be able to direct, guide, and control other team members in line with the principles and practice of the discipline. The importance of a good and knowledgeable facilitator cannot be over-emphasized as the choice can determine the success or failure of the exercise.

Support for the Practice

A major risk in the adoption of value management is the level of support from clients, statutory or regulatory bodies as well as from top management staff of an organisation. The risk may not be pronounced in countries where the disciplined has been embraced and is currently adopted as a project management tool.

However, in countries where most of the construction stakeholders still perceive the practice as another cost-cutting exercise geared towards extorting more money from clients, the outcome and recommendations from the team may even be discarded, rendering the whole process a waste of time and resources with no impact on the project at hand.

Timing of the Exercise

Another risk that can hinder value management is the choice of the right stage of a project at which to implement it. Some schools of thought believe that it should be carried out at the initial phase before the production of final drawings. The problem with this is that there may not be sufficient information for the team to work with and make reasonable suggestions and recommendations. Another school of thought is of the opinion that the exercise should be introduced after the production of detail drawings. There are two issues with this: one is that the project will have to be delayed for at least a week for the exercise to be conducted. The other issue is that recommendations from the exercise will usually lead to amendments and alterations of the existing design which may be at extra cost to the design team. This may eventually lead to a further delay in the actual commencement of the project.

An uncommon and unpopular opinion is for the practice to be introduced as soon as the commencement of site activities. A major base of argument of this view is that before the commencement of a project, there should have been sufficient information for a value management team to work with and that will enhance the success of the exercise. However, this has many disadvantages as the feedback from the exercise will surely slow down the pace of work, may lead to rework, and an eventual waste of resources regarding part of work that would have been completed before feedback from the value management team.

The right timing may depend on projects', clients' and other stakeholders' characteristics. However, regardless of the timing selected for conducting the exercise, the risk must be appropriately evaluated so as not to affect the overall performance of the project. This is further explained in Chap. 8 as one of the principles of sustainable value management.

New Concept for Stakeholders

This risk is concerned with the initial acceptance of the practice of value management by stakeholders, especially members of the original design team for a project. The conclusions and recommendations of value management exercise can alter or change completely the existing proposals made by various members of the design team. This may not be welcomed by some of the members, especially if they perceive it as challenging their professional competence and capability.

Cost of Value Management in Construction

This section explains the various costs associated with a value management exercise. These costs are incurred before, during, and after the exercise. This is one of the challenging areas when convincing clients and other stakeholders to adopt a value management exercise for their project. However, observations from previous projects where value management was carried out and the recommendations adopted, revealed that the cost of a value management exercise is relatively cheap. The cost is low when compared to the added function to the project as a result of the exercise and the eventual savings on the overall project cost. The major costs of conducting a value management exercise are explained as follows.

Cost of Facilitator

A facilitator is an experienced value manager tasked with the responsibilities of planning, controlling, and managing a value management exercise and reporting the outcome to the clients or whoever has commissioned the exercise. It therefore follows that a facilitator is a professional who devotes his or her time to the success of the exercise and as such, must be paid accordingly. The cost may include a professional fee, transport costs, and the like. The payment may depend on several factors including the experience of the facilitator, the type and nature of the project, the mode of engagement, as well as the size and cost of the project. The payment may be made according to an approved scale of fees basis, based on man-hour rate, or any other method which must be agreed upon and approved prior to the commencement of the exercise.

Cost of Other Participants

The participants are other members of the value management apart from the facilitator. These people may or may not be experienced in value management, professionals, experienced in construction, and members of original design team. The identified attributes affect the cost of the individuals and their payment may depend on the same factors as fir the facilitator. The payment method should be decided and settled before the exercise and it can be through any of the remuneration means discussed for the facilitator.

Cost of Venue

A conducive environment with basic facilities and amenities is fundamental to conducting productive meetings. Judging by the expectations from the value management exercise, the venue is expected to be good enough for members of the team to be able to perform their function without any form of hindrance. Depending on factors such as the type of client, the nature of project, and the procedure of the workshop to be adopted, the cost of the venue does not only include the cost of securing or renting the venue but also that of siting the venue in the right environment. If a 40-h procedure is to be adopted, it is better for it to be organised as a retreat in a secluded place, preferably a hotel or guest house with conference and accommodation facilities. This will reduce the stress of team members travelling from one area to the other will also ensure that members can have informal discussions after the official session is over. An alternative is an E-workshop where discussions are held via the Internet, thus eliminating the cost of a venue. However, this comes with the many disadvantages associated with the use of modern day information and telecommunication technologies (ICT) if not properly managed.

Administrative Cost

Apart from the provision of basic social amenities, it is also necessary to allocate cost for administration purposes. These may include insurance, office supplies, wages and benefits of accounting staff, wages and benefits of secretary(s), wages and benefits of legal personnel, among others where applicable.

Information Gathering Cost

A key principle of value management is the comparative analysis of alternatives. This implies that the more alternatives there are and the more detailed information is available about them, the better the ability of the value management team to brainstorm and identify the best alternative, using the principle of function and cost. One of the costs associated with value management is that of collecting information regarding function, cost (initial, annual, running, maintenance), lifespan, and the physical characteristics of elements or components to facilitate deliberation by team members.

Indirect and Other Costs

These are costs that are not directly associated with the actual value management exercise but are necessary in achieving the overall goal of the practice. The indirect cost may include the cost of buying equipment to actualise recommendations emanating from the value management exercise, the costs required in altering or changing the original or existing design to accommodate proposals from the exercise, and the general cost of following up on the understanding and the implementation of the recommendations.

Summary

The concept of value management was introduced to compare alternative materials in order to arrive at the one that provides the best function at the lowest possible overall cost. The chapter revealed that the concept was first introduced to the manufacturing industry but it has since gained popularity in other sectors, including the construction industry. Owing to the varying knowledge and different levels of perceptions of the essence and principles of the operation of the practice, various terms were attributed to the discipline but the most common and all-encompassing one is that of value management. As beneficial as the practice has been in countries and projects where it has been adopted, there are some risks that require attention in order to achieve the purpose for which it was conceived. The direct and indirect costs of conducting the exercise were also identified and discussed in this chapter.

The purpose of explaining value management as a concept in this chapter is to be able to build on the knowledge for subsequent chapters in order to achieve the objective of this book. For instance, Chap. 3 explains value management as a discipline, chapters seven and eight discuss the adoption of the discipline for other construction-related practices while chapters nine, ten, eleven, and twelve build on this to explain the stakeholders, barriers, drivers, and benefits of value management respectively.

References

De Leeuw, C. P. (2006). Value management—The new frontier for the quantity surveyor. Paper presented at the *22nd Biennial Conference/General Meeting on Quantity Surveying*. Abuja: Nigerian Institute of Quantity Surveyors.

Finnigan, A. (2001). *Value engineering. The University of Queensland. Design methods fact.* Retrieved May 12, 2015, from http://www.mech.uq.edu.au/courses/mech4551/.

Green, S. D., & Moss, G. W. (1998). Value management and post-occupancy evaluation: Closing the loop. *Facilities, 16*(1/2), 34–39.

Hewage, K. N., Gannoruwa, A., & Ruwanpura, J. Y. (2011). Current status of factors leading to team performance of on-site construction professionals in Alberta building construction projects. *Canadian Journal of Civil Engineering, 38*(2011), 679–689.

Jay, C. I., & Bowen, P. I. (2015). Value management and innovation: A historical perspective and review of the evidence. *Journal of Engineering, Design and Technology, 13*(1), 123–143.

Karunasena, G., Rathnayake, R. M., & Senarathne, D. (2016). Integrating sustainability concepts and value planning for sustainable construction. *Built Environment Project and Asset Management, 6*(2), 125–138.

Kelly, J., & Male, S. (2006). Value management. In J. Kelly, R. Morledge, & S. Wikinson (Eds.), *Best value in construction* (pp. 77–99). London: Blackwell.

Kelly, J., Male, S., & Graham, D. (2014). *Value management of construction projects*. Oxford: Wiley.

Leung, M.-Y., & Yu, J. (2014). Value methodology in public engagement for construction development projects. *Built Environment Project and Asset Management, 4*(1), 55–70.

Liu, G., & Shen, Q. (2005). Value management in China: Current state and future prospect. *Management Decision, 43*(4), 603–610.

Male, S. (2002). A re-appraisal of value methodologies in construction. *Construction Management and Economics, 11*(2002), 57–75.

Miles, L. D. (1972). *Techniques for value analysis and engineering* (2nd ed.). New York: McGraw-Hill.

Oke, A. E. & Ogunsemi, D. R. (2011). Value management in the Nigerian construction industry: Militating factors and the perceived benefits. In *Proceedings of the Second International Conference on Advances in Engineering and Technology* (pp. 353–359). Faculty of Technology, Makerere University, Uganda.

Oke, A. E., & Ogunsemi, D. R. (2013). Key competencies of value managers in Lagos state, Nigeria. In S. Larry & S. Agyepong (Eds.), *Proceedings of 5th West Africa Built Environment Research (WABER) Conference* (pp. 773–778). Ghana: Accra.

Oke, A. E., Aghimien, D. O., & Olatunji, S. O. (2015). Implementation of value management as an economic sustainability tool for building construction in Nigeria. *International Journal of Managing Value and Supply Chains, 6*(4), 55–64.

Palmer, A., Kelly, J. & Male, S. (1996). Holistic appraisal of value engineering in construction in United States. *Construction Engineering and Management*, 324–326.

Perera, S., Hayles, C. S., & Kerlin, S. (2011). An analysis of value management in practice: The case of Northern Ireland's construction industry. *Journal of Financial Management of Property and Construction, 16*(2), 94–110.

Shen, G. Q., & Yu, A. T. (2012). Value management: Recent developments and way forward. *Construction Innovation, 12*(3), 264–271.

Shen, Q., & Liu, G. (2004). Applications of value management in the construction industry in China. *Engineering, Construction and Architectural Management, 11*(1), 9–19.

Short, C. A., Barett, P., Dye, A., & Sutrisana, M. (2008). Impacts of value engineering on five capital arts projects. *Construction Management and Economics, 35*(3), 287–315.

Society of American Value Engineers. (2015). *What is value engineering?* Retrieved May 27, 2016, from http://www.value-eng.org/.

The College of Estate Management. (1995). *Value engineering*. Retrieved May12, 2016, from http://www.cem.ac.uk/postalcourses.

The Institute of Value Management. (2015). *What is value management?* Retrieved May 12, 2016, from http://www.ivm.org.uk/what_vm.htm.

Yekini, A. A., Bello, S. K. & Olaiya, K. A. (2015). Application of value engineering techniques in sustainable product and service design. *Science and Engineering Perspectives, 10*(Sept), 120–130.

Chapter 3
Value Management as a Discipline

Abstract The idea behind the introduction, practice, and adoption of value management is for it to serve as a method of analysing, enhancing, and improving acceptable standards of controlling and managing elements and components of products or projects from inception to completion, commissioning, use, re-use, and eventual demolition. In explaining value management as a discipline, this chapter describes various procedures necessary in the conduct of the discipline, different means of planning and organising it, the concept of function analysis, issues relating to whole-life costing, distinguishing between quality and function of elements, the principle of value, cost, worth, and price as well as the constituents and components of unnecessary cost.

Keywords Function analysis · Team composition · Unnecessary cost · Value · Value management · Whole-life costing

Introduction

The idea behind the introduction, practice, and adoption of value management is for it to serve as a method of analysing, enhancing, and improving acceptable standards of controlling and managing elements and components of products or projects from inception to completion, commissioning, use, re-use, and eventual demolition. This is with the view to arriving at the elements that provide the best function in relation to adopted measures of function at the overall least possible cost. The essence of a discipline is to ensure proper procedure and standards for individuals or processes and this underscores the concept of value management.

In explaining value management as a discipline, this chapter describes various procedures necessary in the conduct of the discipline; different means of planning and organising it; the concept of function analysis; issues relating to whole-life costing; distinguishing between quality and function of elements; the principle of value, cost, worth and price; as well as the constituents and components of unnecessary cost.

A.E. Oke and C.O. Aigbavboa, *Sustainable Value Management for Construction Projects*, DOI 10.1007/978-3-319-54151-8_3

Value Management Procedure

Value management adopts a systematic approach and various orderly stages are followed in conducting the exercise. The procedure for a typical value management activity is described in this section based on existing reports and studies (Miles 1972; Palmer et al. 1996; Finnigan 2001; Male 2002; De Leeuw 2006; Kelly and Male 2006; Short et al. 2008; Shen and Yu 2012; Kelly et al. 2014; Jay and Bowen 2015). They include the following:

- Pre-workshop (or study) phase
- Information phase
- Analysis of function phase
- Creative phase
- Evaluation phase
- Development phase
- Presentation phase
- Post-workshop (or study) phase

Pre-study Phase

This is the first phase of a value management exercise which is concerned with the introduction of various aspects of the work, the team members and other necessary information for the purpose of conducting an effective value management exercise. It is concerned with orientation meetings; team members and structure; duration, condition, and location of the study; site visit; information gathering; and a verification of cost estimate. According to Kam Shadan (2012), the pre-study phase involves planning of the value management study, while highlighting the following as essential:

- Obtaining project documents;
- Verifying value management schedules and agenda;
- Suggesting a format for designer presentation;
- Outlining project responsibilities;
- Establishing the owner's performance and acceptance requirements;
- Conducting a coordination meeting and
- Identifying project constraints.

The first activity is the orientation meeting between clients, owners, sponsors, the original design team, whoever is commissioning the value management study, members of the value management team, and others who have an interest in the project. However, it should be noted that it is possible for team members' selection to be in progress at this stage but at least one member of the team, possibly the facilitator, should have been identified and possibly commissioned for the purpose of the value management exercise. The objective of the orientation meeting is to officially brief the value management team members about the project and charge them with their expected responsibilities concerning the project under consideration. The outcome of the orientation meeting is considered important in finalising the value management team structure.

Every co-opted team member has some special attributes, skills, competence or characteristics that can contribute positively or negatively to the structure of any team in which they are involved. The second stage of the pre-study phase of the value management exercise is finalising the team structure which is concerned with the composition of the team. It is the responsibility of the facilitator to decide the configuration of the team, judging from some of the characteristics of the project and the expected outcomes. This stage is related to the actual selection of the team members.

Value management is a team-based management system and the intention is that every member of the team has one or more good ideas to contribute towards enhancing the functions of various elements using alternative materials at the lowest possible cost. The choice at this stage depends mostly on the professions of the members but attention should also be paid to the integration of some of the original project design team. The individual roles of the team members also need to be considered. The number of members required for the team should also be decided at this stage: there should not be too many or too few value management team members for the purpose of the exercise to be achieved. However, it is preferable to have an odd number rather than an even one in case there is a need to put a decision to the vote. For the professionals to be selected, the facilitator needs to ensure that several factors such as value management experience, professional competence and experience, the nature of the project, and the complexity of the project are the basis for their selection. Another aspect is the incorporation of some of design team members of the project under consideration. It is not possible to include all of them; the onus is therefore on the facilitator to decide which of them will contribute more meaningfully to the value management team. Team roles of the individuals who will form the value management team should also be considered as an imbalance or absence of some important ones may affect the smooth progress of the team.

Another important decision is the determination of the structure as well as the duration of the value management study. Depending on the structure to be adopted, the length of the study should be specified, taking into consideration factors such as the complexity of the project, the cost of project, and the type and nature of the team members.

At pre-study phase, the location of the study as well as its conditions should be determined. The choice of location will depend largely on the availability of funds and the nature of the project under consideration. Some scholars have suggested the use of a project site meeting office to avoid spending extra money on renting a venue but this comes with its advantages and disadvantages. To avoid distractions, it is better if the location is a secluded area with only the members and supporting staff in attendance. It is also possible to adopt an electronic conference system but the conditions should be well stated and understood by team members in order to avoid absenteeism and lateness. This may also be a challenge in countries where there is persistent problem with Internet connectivity and other ICT infrastructure. It is the responsibility of the facilitator in conjunction with the others members to determine the venue of the study as well as the conditions that will guide the team activities.

At the orientation stage, some vital information is provided but it is also necessary to seek and gather more information about the project. This information may be related to the projects and stakeholders of the project. The project's information is the most important factor with the emphasis on the type, nature, purpose, and the like. The next stage is to visit the project site with the objective of gathering more information about the project and familiarizing everyone with the project environment. Also, there is a need to verify the cost estimate and other existing information and documents regarding the project.

Information Stage

This is the stage of the value management study where necessary, relevant, and correct information concerning the project is gathered from appropriate source(s). The information should be related to various parts, elements, components, or the whole project and should be sourced directly from clients or their representatives. It is necessary at this stage to gather and record all the functions, whether they are primary or secondary, with the emphasis on the needs and wants of the client, the project restrictions, the limit of expenditure, and the time available for design, construction, and usage. According to Kam Shadan (2012), the information phase involves preparing for the workshop which entails the following:

- Distributing documents to team members, that is, drawings, specifications, cost estimates, design criteria, site conditions as well as utilities, operation and maintenance issues and
- Familiarizing team members with the project.

For every project conceived by an individual, there is a particular need in the mind of the client. At the conception of the project, the client's *needs* are the major requirements that underscore the reason for the conception of the project. Related to these are the *wants* of the client which do not serve the actual idea at the conception of the project but are additional. These should also be considered at the information gathering stage.

Another aspect of information gathering stage is the identification of various constraints associated with the project under consideration. The constraints may be related to statutory requirements, infrastructure planning requirements, general building regulations, design requirements for projects in the area, design requirements for the type of project in consideration, the nature of the soil and other site conditions, the location of the project in respect to other projects, the topography of the site, and similar considerations.

The limit of amount of money and resources that are available for the initial, running, and final costs of the project should also be ascertained at the information stage of the project. This is known as budgetary limit and it is concerned with the overall whole-life costing of the elements and components of the project.

The other information to be obtained and sourced from clients, clients' representatives or any other appropriate individual is the period of time available for the project. The time span is at three stages. The first and initial is the time required for the various forms of designs and associated practices which may include sketches, scheme design, detail design and associated activities such as a feasibility study and viability appraisal. The duration required and the time available for the actual construction of the project (work on site) should also be ascertained while the last-time related information is the period of use, possession, or ownership of the project. This is important in determining the return on the client's investment which is a key principle of value management.

Some schools of thought believe that the information phase is in two stages, namely, presentation and function analysis. Owing to the importance of analysing the significance of these to value management, the issues relating to the presentation phase are discussed under the information stage while the function analysis is treated as an important stage of the process.

Function Analysis

An analysis of function is fundamental to the principle and process of value management and the emphasis of the technique is on what the elements or components are doing rather than what they are. Function is a set of attributes, characteristics, or features that make an element, part, item, or component to work in a particular way. The function of an element, component, or part of an infrastructure should be identified and recorded accordingly. However, it should be noted that parts of the project or the entire project may serve more than one function. In view of this, the function should be identified as either primary (basic) or secondary, judging by the needs and wants of the client, the purpose of project, and the statutory purpose of the parts.

The following questions have been suggested in the examination of functions in value management:

- What is it?
- What does it do?
- What else does it do?
- What else can it do?
- What does it cost?
- What is its value?

The first question relates to what the element under consideration is all about in terms of the wrong perceptions of members of the value management team. This can be tackled by enquiring from members what their perceptions of the element are or by highlighting what the element is not meant for, especially in relation to the project. It is essential for the team members to understand and agree on the specific purpose of elements or components for a better analysis of their function.

For the second question on what the element does, it is necessary to highlight and understand the basic and primary function of various identified elements. This purpose or function should be outlined in a clear and unambiguous manner for easy understanding. For instance, a verb and a noun can be paired for the identification of the function of elements: this may include such statements as 'retain heat', 'absorb sound', 'beautify project', and 'support load'.

The next issue of what else the element does relates to the secondary function of the elements. These are not the primary purpose of the element but they can serve these functions without compromising their primary roles. For instance, the primary function of a column located in the front view of a building may be to support load but it can also serve the purpose of aesthetics as secondary function. This can be achieved through gathering information about the elements and various functions they have performed for other construction projects at one time or the other. But, most importantly, the peculiarities and distinctiveness of the present project that is being 'value managed' should be the point of reference in defining function.

The other question of what else can an element can do is related to what else does it do, except that the former functions have to be identified in relation to the project and most frequently, against the norms and existing knowledge. Determining what else an element does will be better achieved by allowing members of the value management team, especially the non-professionals or those without expertise in the field of the element in consideration, to come up with their own suggestions and ideas.

The issue of cost is an important element of function analysis. The cost of each element should be identified and discounted accordingly. Various costs (as discussed under the whole-life costing section of this chapter) should be considered for each of the elements and this requires the expertise of an estimator, quantity surveyor, or cost engineer.

The last aspect of function analysis is the establishment of the value of each of the elements of the project. Value is directly related to function and inversely proportional to cost, that is: *Value = Function/Cost*. This implies that the value of element cannot be established unless and until all the necessary functions and associated costs are identified and evaluated accordingly. The concept of value in relation to other related terms is discussed further in designated sections of this chapter.

Function analysis is a major principle of value management: it emphasized function rather than just cost which distinguish it from other cost-control, cost-saving, and cost- cutting exercises in which the emphasis is strictly on cost of project. The analysis of function ensures that the basic needs of the client in term of safety, reliability, quality, and aesthetics are not compromised. According to Kam

Shadan (2012), preparation of the cost, whole-life costing and energy models are carried out at this stage. Other activities at this stage of a typical VM study include the following:

- Reviewing cost estimate;
- Preparing cost models;
- Preparing Pareto models where necessary;
- Identifying high cost areas;
- Preparing life-cycle costs;
- Performing function analysis;
- Calculating cost/worth of each function; and
- Identifying areas of high cost or low value.

Kelly et al. (2014) identify four levels in functional analysis, namely task, space, elements, and built form. The task phase is concerned with the way the client, owner, sponsor, financier and other principal parties to the project initially perceive or view a particular task or problem. This will be affected by their interest, training, profession, and experience. The second level, the space stage, is concerned with the involvement of design team members such as the architect, engineer, land surveyor, quantity surveyor, construction manager, and project manager in the interpretation, analysis and evaluation of the task for the purpose of preparing the brief. The brief is expected to be clear and easy to understand by every member of the team, including the non-construction professionals. The third level is the elemental stage where information from the task and space level is used to produce a structural form of the project for ease of interpretation. The last level is when the elements in level three are combined and framed to assume an identity of the built form of the project.

Another means of analysing function is through the use of a function logic diagram by constructing a function analysis system technique (FAST). The diagram resembles a decision tree and helps to answer the questions 'Why?' and 'How?' in order to identify the real functions rather than symptoms.

Creative Phase

This is a follow-up to the information gathering and function analysis phase: it is also known as the speculation phase. The essence of this phase is to arrive at a pool of ideas and suggestions on how to address the functions of components, elements, or features that have been identified earlier in the course of the study. The suggestions are without restrictions and various forms of creative methods and thinking can be adopted. These include brainstorming, the Synectics method as well as the Gordon approach. The basic activities attributed to this stage, according to Kam Shadan (2012), include the following:

- Generating a list of ideas to meet required functions;
- Doing creative thinking;

- Practising group thinking; and
- Examining project functionality.

Judging by the categories and diverse individuals that form the value management team, it is possible that some ideas may seem unreasonable and odd but this stage it is more about the quantity of ideas rather than the quality. In fact, previous studies have shown that some of the ideas thought to be strange, weird, and useless often become the best ones after evaluation by the team members. This indicates that all ideas are welcome from team members and are recorded accordingly without favouritism and restraint.

Evaluation Phase

The creative phase is designed to generate ideas of various sorts while at the evaluation phase, the ideas are assessed, examined, and evaluated to arrive at the valuable and superior ones (quality ideas). This implies that various ideas must have been generated for this stage to be effective and achieve its purpose of reducing and filtering them to a manageable number for further evaluation and development.

For the ideas to evaluated, it is necessary to set criteria for their acceptance and rejection to avoid the perception of being biased. The criteria used vary from one project to the other and may also depend on other characteristics such as the client factor and statutory regulations. For value management to be effective, the most important factor should be related to value which encompasses cost (initial, annual, running), function (health and safety, quality, security, aesthetics), time relatedness (reliability, durability, lifespan) and any other general criteria as agreed by the team members. It is necessary to evaluate each of the ideas against functional requirements in order to screen out less promising ideas (Kam Shadan 2012). The process followed in the evaluation stage may include the following:

- Ranking ideas;
- Listing the advantages of ideas;
- Listing the disadvantages of ideas;
- Identifying potential for acceptance of an idea; and
- Developing the best ideas.

The final decision usually rests with the facilitator but it is not uncommon for ideas to be put to the vote when there seems to be no consensus among team members. Then the facilitator as the leader of the team may need to adopt one or more leadership styles including charismatic, transformational, democratic, autocratic, consensus, and shareholders and the choice and application will depend on several factors that are peculiar to the project and value management principles. However, it is the responsibility of the facilitator to understand and know what method is to be adopted at various times.

Development Phase

The selected and adopted ideas go through further development through detailed investigation with the emphasis on the feasibility and viability of the ideas. The feasibility aspect is concerned with technical issues and the possibilities of the ideas while viability is related to the financial and economic aspect of the idea. The idea of whole-life costing comes into play at this stage using various project evaluation and investment appraisal techniques. It is important to consider the ideas from all angles of value as described earlier.

The development phase is an important stage of a value management exercise and care should be taken to ensure that ideas that will be recommended in the presentation phase are carefully examined and considered. The ideas should be in line with the value management principle of value with the emphasis on associated function and cost. The ideas should be evaluated along with the existing ideas of the project and those found to have less value should be rejected.

Presentation Phase

After developing the ideas and arriving at the ones that provide best value and return on investment, the ideas are further processed and refined with the aid of pictorial representations or drawings, and cost information for better understanding, even to a layman.

This is the feedback and handover stage. Information and recommendations from the value management team are presented to the commissioned agents who can be clients, the design team or other bodies. Regardless of who commissioned the value management study, it is essential for the original design team of the project to be included in the presentation phase for proper understanding and ease of application of recommendations from the value management team. Some authors separate the feedback from the presentation as the last phase of value management but it is believed that the essence of the presentation of the ideas emanating from value management study is to handover such recommendations to the body that commissioned the study as a form of feedback.

Post-study Phase

This phase is not directly link to the process of value management but forms the aftermath of the exercise. It is a review exercise and it is usually undertaken after the presentation, handover, and feedback of recommendations emanating from the study. At this phase, the report of the exercise and the review are presented only for the members who carried out the value management exercise. The report is not the same as that presented to the body that commissioned the study at the presentation phase but rather it is a review report that may detail the activities of individuals during the value management exercise; lesson learnt from individuals, process and the project; ways for improvement, and other important details. The essence of this is not only to help members of the team achieve better performance in subsequent value management studies but also to provide historical data and a basis for individuals with an interest in the study as well providing basic information for future value management studies.

Another area of concern of the post-study phase is the monitoring of the implementation of the recommendations that emanated from the value management exercise to the under consideration. It will be fruitless if the suggestions from the exercise are only partially implemented or completely discarded. In fact, it will amount to a waste of money, time, and other resources if the recommendations are poorly implemented by the project design team. It is therefore the responsibility of the value management team, together with the construction team members, to ensure a smooth implementation of their recommendations regarding the project. This may involve the clarification of grey areas by the facilitator or by a selected member of the team. It is believed that this is one of the reasons for the involvement of members of the original design team in the value management study.

Similar to the implementation phase is the follow-up aspect which is concerned with collating a record of how beneficial the recommendations of the value management study are to the project. For policy formulation and recommendations to clients of other projects, the follow-up process of value management is designed to gather information relating to the cost saving on the project, satisfaction of stakeholders with the implementation and effect of the study on the project, the level of improvement in the overall project as a result of the study, a record relating to differences between the original and actual duration as a result of the study, and other performance-related measures.

Facilitation of Value Management

Since time value management was introduced to the manufacturing industry, several strategies have been adopted to conduct the exercise without jeopardizing the basic purpose and principles of the study. This section describes various methods of facilitating value management workshops which include the following:

- 40-h workshop;
- 24-h workshop and other shortened study;
- Job plan;

- Orientation meeting;
- Value management edit;
- Electronic method through virtual team, video conferencing;
- Delphi approach;
- Contractor's change proposal;
- Group Support System; and
- Concurrent study.

The list is not exhaustive as various means of facilitating the study are emerging based on the perspective, opinion, experience, profession, and environment of operation of value management practitioners and researchers. The 40-h workshop is still the most common and widely accepted and adopted technique. However, regardless of the technique of facilitation adopted, it is necessary that adequate time is given to the various phases of the exercise and that individual members of the team are focused not only on contributing their ideas but are also interested in the overall success of the study.

Roles of Value Management Team Members

Value management is a team-based study that does not only rely on the professional qualifications or experience of team members but also on their teamwork abilities. Because of the involvement of individuals in value management team, an important issue in managing and fostering cohesion in a team is to ensure that individuals with appropriate team roles are selected. There are nine known and well researched team roles that are required for team success.

The team roles, as well as their requirements at different stages of development, are discussed in detail in Chap. 4 of this book under the 'team roles of project stakeholders' section. Moreover, team roles for sustainable value management are also discussed in Chap. 9. However, the team roles expected in a value management exercise include the following:

- Plant;
- Specialist;
- Completer or finisher;
- Implementer;
- Team worker;
- Monitor or evaluator;
- Shaper;
- Coordinator; and
- Resource investigator.

Whole-Life Costing

There are various costs associated with an item, element, component, part of structure as well as the structures themselves. The most common is the initial capital which is the original cost of acquiring or putting the item in place. However, there are other costs associated with every item that is required to be discounted after a particular period and this is necessary and useful for a comparative analysis of alternative items. Life cost, life-cycle cost, or whole-life cost is the summation of total costs of the initial and other operating costs of an item, element, component, or structure. The cost includes the initial capital, cost of construction or putting in place, running or operating cost, maintenance cost, cost of re-use, cost of demolition, as well as cost of eventual disposal.

The constituents of whole-life costing may include the following, depending on the kind of item under consideration:

- Cost associated with preparatory works;
- Initial capital cost;
- Professional fee;
- Site cost;
- Annual cost;
- Insurance;
- Legal cost;
- Periodic cost;
- Other running or operation cost;
- Replacement cost;
- Maintenance cost;
- Alteration cost;
- Service cost;
- Energy cost;
- Taxation cost;
- Commissioning cost;
- Demolition cost;
- Cost of re-use or salvage;
- Cost of conversion;
- Disposal cost;
- Other client or associated costs.

There are various methods for discounting various forms of cost associated with an element or project to the same basis. The methods are derived from the basic compound interest formulae that are commonly use for cost-in-use calculations. They include, but are not limited to, the following:

- Net present value (NPV)
- Return on capital employed (ROCE)
- Annual equivalent value (AEV)
- Discounted cash flow (DSF)
- Internal rate of return (IRR).

Value, Function and Quality

Value management is designed around the principle of value, function, and quality but the main emphasis is on value which relates to other indices as well as function and quality. The value of a product, element, component, or project is a balance of the function, quality, cost, time, and other measures of project delivery and performance. According to Norton and McElligott (1995), the three major ways to improve value from a cost perspective which usually form the basis of value management study are highlighted below:

- Same functions at lower cost;
- Additional function at same cost; and
- Additional function at lower cost.

The first means is to retain or provide a function of elements, components, parts, or entirety of project at a lower cost to the existing cost as quoted by the estimator or quantity surveyor of the original project design team. Another approach to value improvement is the provision of additional function to the parts or the entire project without any increase in the initial or budgeted cost. The third approach is concerned with increased or additional function of the projects in part or as a whole while reducing cost. Value encompasses other measures of project performance and value management should be conducted with the focus on quality, health and safety, cost, function, aesthetics, reliability, and overall duration of the structure.

Value, Cost, Worth and Price

Judging from the principles of value management, value as a term encompasses other related items including cost, worth, and price. The principle of value in value management using the concept of function does not only consider the worth at the lowest cost but various prices associated with an element or component including initial, running, and final costs. The distinction between price and cost is vigorously debated by various authors who explain the two terms from various views and fields of endeavours. They are related and the best way to address them is based on the common phrase that "…one man's price is another man's cost". However, various economics books and research material can be consulted for further explanation of these terms.

Concept of Unnecessary Cost

Unnecessary costs are costs associated with an elements or components that do not contribute any value to the project. It is the cost of an unnecessary material, feature, process, or property which does not provide any performance-related functions such as appearance, reliability, or customers' satisfaction. It can also be the cost of a function of an element or project that is not relevant to the client' needs or does not contribute to the purpose of the project under consideration. The causes of unnecessary cost may include the following:

- Paucity of cost data;
- Lack of details about client's needs and user's requirement;
- Lack of life cycle cost estimate;
- Politically related issues;
- Lack of experience of project team members;
- Lack of capital and funds for necessary preliminary study;
- Ambiguous goals, scope and objectives of project;
- Incomplete project design and lack of other related documents;
- Usage of unsuitable standards or lack of specifications;
- Lack of effective communication among project team stakeholders;
- Continuous changes of owner's needs and requirements;
- Unrealistic and impractical designs;
- Wrong perception of costs and individual attitudes;
- Resistance to change by stakeholders;
- Wrong information, especially in contract documents;
- Conflict and disputes due to unresolved and unclear issues in project documents;
- Legal battle as a result of the project processes or error in contract documents; and
- Hasty decision based on false and unproven assumptions and opinions.

Summary

The objective of this research book is to further existing knowledge in the area of value management in the construction industry and on how sustainability issues can be factored into construction project execution. The discipline of value management was explained in this chapter to provide a background for the discussion relating to its application for sustainable construction. To further this explanation, Chap. 7 of this book discusses the adoption of value management for various construction management approaches while Chap. 8 focuses on the achievement of sustainable construction through the application and adoption of value management. Each of the value management stages are discussed in Chap. 8 with the emphasis on incorporating construction sustainability issues into the study.

References

De Leeuw, C. P. (2006). Value management—The new frontier for the quantity surveyor. Paper presented at the *22nd Biennial Conference/General Meeting on Quantity Surveying*. Abuja: Nigerian Institute of Quantity Surveyors.

Finnigan, A. (2001). *Value engineering. The University of Queensland. Design methods fact.* Retrieved May 12, 2015, from http://www.mech.uq.edu.au/courses/mech4551/.

Jay, C. I., & Bowen, P. I. (2015). Value management and innovation: A historical perspective and review of the evidence. *Journal of Engineering, Design and Technology, 13*(1), 123–143.

Kam Shadan, P. E. (2012). *Construction project management handbook.* Washington: Federal Transit Administration. Retrieved September 14, 2016, from https://www.transit.dot.gov/sites/fta.dot.gov/files/FTA_Report_No._0015_0.pdf.

Kelly, J., & Male, S. (2006). Value management. In J. Kelly, R. Morledge, & S. Wikinson (Eds.), *Best value in construction* (pp. 77–99). London: Blackwell.

Kelly, J., Male, S., & Graham, D. (2014). *Value management of construction projects*. Oxford: Wiley.

Male, S. (2002). A re-appraisal of value methodologies in construction. *Construction Management and Economics, 11*(2002), 57–75.

Miles, L. D. (1972). *Techniques for value analysis and engineering* (2nd ed.). New York: McGraw-Hill.

Norton, B. R., & McElligott, W. C. (1995). *Sustainability of construction materials*. London: Macmillan Building and Surveying series.

Palmer, A., Kelly, J. & Male, S. (1996). Holistic appraisal of value engineering in construction in United States. *Construction Engineering and Management*, 324–326.

Shen, G. Q., & Yu, A. T. (2012). Value management: Recent developments and way forward. *Construction Innovation, 12*(3), 264–271.

Short, C. A., Barett, P., Dye, A., & Sutrisana, M. (2008). Impacts of value engineering on five capital arts projects. *Construction Management and Economics, 35*(3), 287–315.

Part III
Sustainable Construction

Chapter 4
Construction Projects and Stakeholders

Abstract Owing to new and emerging performance measurement for construction projects, clients of construction projects will continue to demand more complex projects and it is the responsibility of the participants, especially the contractors and consultants, to ensure that the industry continues to be relevant in the face of changing demands, needs and requirements. One of the means of achieving this is through adhering to the construction process and ensuring that the right mix of team members in terms of team roles and profession is in place for all construction work. In view of this, it is deemed necessary to discuss the nature and concept of construction projects in term of their objectives and processes. Various participants and stakeholders responsible for construction activities are explained while team roles as they relate to the key stakeholders are also discussed. In addition, the challenges of the construction process are highlighted and explained with a view to adopting the principle of value management to reduce or eliminate these.

Keywords Construction project · Project objective · Project stakeholder · Project success · Team role

Introduction

The focus of this research book is the application of value management in order to achieve sustainability in construction projects. In view of this, it is necessary to discuss the nature and concept of construction projects in terms of their objectives and processes. Various participants and stakeholders responsible for construction activities are explained while team roles as they relate to the key stakeholders were also discussed. The challenges of the construction process are also highlighted and explained with a view to adopting the principle of value management to reduce or eliminate these. Construction projects in this book are synonymous with architectural, engineering, and construction (AEC) projects, terms which have been used interchangeably by authors, experts and practitioners.

A.E. Oke and C.O. Aigbavboa, *Sustainable Value Management for Construction Projects*, DOI 10.1007/978-3-319-54151-8_4

Objectives of Construction Projects

Construction projects are capital project and a major engine of economic growth. They consist of multi-disciplinary individuals and interrelated work activities operating within a confined schedule, budget and scope, or standard (United Kingdom. Cabinet Office 2011; Takim and Akintoye 2002; Banaitiene and Banaitis 2012; Kam Shadan 2012). For instance, scholars have found that the construction industry contributes about 7% to the GDP of the UK economy (United Kingdom. Cabinet Office 2011). However, owing to vague and increasing complex expectations of clients, complicated organisational structures, the dynamic economic situation, insufficient substructure information, among others, construction projects and their associated activities normally involve uncertainty and risks (Leung et al. 2010).

The objective of construction projects depends on the perspective of the individuals and stakeholders. The main goal of all forms of contractors of construction projects—principal, trade or subcontractor (nominated or domestic)—are to deliver the project successfully in consideration of the commissioned agents' requirements—which may be client, owner, or their representative—to acceptable standard, expected cost, specified duration, safety, and other performance measures in return for reasonable profit (The Associated General Contractors of America 2003). To a construction company, the following, among others, may serve as guide in understanding and evaluating whether a project meets the company's strategic objectives and business mission:

- Project profit maximization;
- Consistency of project's scope with company expertise;
- Acceptability of project's risk and uncertainty;
- Satisfaction of stakeholders' need by executing the project;
- Improvement of company's image by the project;
- Maximization of project plant and equipment utilization;
- Project enabling company to enter new market;
- Maximization of project workforce utilization; and
- Project maintaining market position.

To clients, the above identified variables are important but the most important one is value for money in terms of cost or associated benefits, and return on investment. This should also be the concern of other participants engaged by the clients, including the professionals, contractors and others with indirect links to the actual construction process. Others participants are discussed in this chapter under 'stakeholders in project construction' and judging by their role and interests, each one of them is expected to value their company's objectives in addition to the client's goal.

Construction as a Process

Regardless of the type of developmental project, a construction project usually follows the same process, except that some of the stages may be more complex or less feasible in one type of project than the other. The general process of construction of infrastructures can be classified into pre-construction, construction, and post-construction activities. For a typical construction project, the basic process includes conception; inception; feasibility; outline of proposal; scheme design; detailed design; production information; tender action; project planning; site operation; completion; handling over and feedback; operation and usage; demolition and re-use. However, some of the activities or processes can be combined or expanded, depending on the client and project characteristics (United Kingdom. HM Treasury 2014; Kam Shadan 2012).

Conception Phase

This is an important pre-construction stage when the idea of the project is conceived by the client as a result of his desire or need, or sometimes due to the expectation or need of other people, community or the general environment. At this stage, it is possible that the idea may be vague and sketchy but the most important consideration is to have a concept of what is to be done with every detail possible.

Inception Phase

It is possible that the client is not experienced in construction activities and even if the client does have some experience, it will probably be limited. The essence of this stage is to define the requirements of the client through the engagement of a prime consultant, usually a construction project manager, architect, engineer, project manager, construction manager, quantity surveyor, or any other related professional, depending on the nature of project idea and the knowledge and understanding of the client. At this stage it is expected that the idea of the client should be refined through the engagement of professionals who are relevant to the project(s) under consideration for the better understanding of other stakeholders.

The prime consultant, in conjunction with the client depending on the nature of contract between them, is expected to commission other relevant stakeholders, including consultants and professionals, to work further on the client's idea. According to the Isle of Man Treasury Government Office (2002), the essential items of client's brief that should be further developed in the course of construction activities include the following:

- Background information, especially that relating to the purpose of the project;
- Content and scope of project;
- Expected budget constraints;
- Required duration for the project;
- Other required project performance criteria;
- Impact of failure to meet cost, time and other performance target;
- Planning regulations and rules;
- Social brief concerning how the project will be executed as well as stakeholders that will be involved;
- Schedule of required accommodation; and
- Statement of expected project function, process and activities.

Feasibility Phase

This is also known as the feasibility and reliability stage where the possibility and profitability of the idea are examined and evaluated to determine its value and worth in the present and future. Various aspects such as cost, value, function, clients' needs, source of finance, and form of possession and ownership are expected to be established at this stage in order to determine the form the project will take. The cost limit is also set based on the budget or limit of funds available for the project from the client, owner or financier, through approximate estimate principle or based on knowledge of past but similar projects.

The major purpose of a feasibility and viability study for developmental projects includes the following:

- To determine the possibility for return on investment by client, owner, sponsor or financier;
- To evaluate the feasibility of the project providing value for money to be expended;
- To aid decision making of the stakeholders on whether to reject, modify or accept and execute the projects;
- To help to convince investors to support the project in a case where external financier or sponsor are required:
- To serve as a benchmark for other phases in ensuring that the objective and set target are abided with during the course of the project execution; and
- To serve as historical information to guide subsequent projects especially for clients and stakeholders with little experience of the type of project or area where the project is or will be situated.

The feasibility and viability of a project are usually evaluated using the six variables of economic, physical, technological, legal, socio-political, and financial measures. The physical aspect is concerned with the situation of the area and the suitability of the proposed site for the project. Such factors such as the type of projects the area is designed for, the strength of the soil, and the topography of the land/ground are considered for this aspect of the study.

Based on the existing expertise and technological knowledge of construction experts in the area or globally, as the case may be, technological feasibility is concerned with the evaluation of the buildability and construction ability of the project under consideration. There is nothing wrong in trying out new concepts and ideas but they should be feasible within the available technological resources. For sites, such aspects such as a traffic study, zoning study, soil test, water test, and geotechnical investigations are important considerations.

The legality of the project should be examined so that it does not contradict existing laws and regulations of the local, regional, or national government. Information to be sought for legal feasibility includes the planning laws of the area, ownership status of the land or area, laws relating to the transfer of land in the area, and regulations relating to the approval of plan and project.

Depending on the nature of the project, the clients and users for whom the project is designed, the issue of socio-political feasibility cannot be overemphasized. The level of involvement and participation of the members of the communities hosting the project should be determined and assessed to foster a good relationship for the success of the project. The rules and regulations for this aspect may be written or unspecified but it is the duty of the project team to seek information regarding social issues relevant to the area. The political aspect of a typical feasibility study is concerned with diplomatic issues prevalent in the country and the area, especially for foreigners. The potential impact and effect of such factors including war, change of power, and protests should be considered for an effective feasibility and viability study.

The economic feasibility is related to fiscal and monetary policy as well as the prevailing financial conditions of the expected customer, buyer or user in the local, national, and international environment where the project is to be located. The issue of tax, inflation, and interest rates are also to be considered for this aspect of a feasibility study. Whole-life costing and investment appraisal systems are usually adopted for effective economic appraisal.

The source of funds for the project should also be determined and evaluated as this forms a major issue for a developmental project, especially those which are complex and capital intensive. The financial feasibility deals with these issues with the emphasis on whether the client or owner has the capital to finance the project through personal money, loans, or other means. Alternatively, there will be requirements to involve investors in form of financier(s) or sponsor(s). The funds in this regard are not only limited to the initial capital but the overall cost of completing the project and in some cases, the running, conversion, and eventual demolition of the project after its lifespan. Financial feasibility is not the same as economic feasibility although both form the major part of a viability study. The former examines the monetary situation or capital required and expected at the start of the project while the latter is more concerned with the economic importance of the project which usually takes effect from the beginning to the completion and close-out of the project.

A typical feasibility report incorporating the six phases explained earlier may be arranged in five sections. As exhaustive as this may be, the nature of the project and other factors may necessitate more sections or further dividing of any of the identified sections. The five sections are the following:

- Introduction and project background;
- Market survey and data analysis;
- Technical feasibility;
- Viability report; and
- Conclusion and recommendations.

Outline Proposal Phase

Depending on the decision at the feasibility stage, the duty of the design team at this stage will be to further develop the idea of the client through various means of presentation. The outline proposal is a brief design stage where the project brief is developed, usually by the architect, engineer, project manager, or prime consultant. The outline proposal report regarding general layout, various forms of designs, and other issues relating to the project construction is prepared for the approval of the client or any other body that may have commissioned the project.

Also at this stage, cost analyses of alternative designs of the project are also prepared using such methods as elemental and comparative cost-planning techniques. The objective is to have various forms of designs and supporting information to support the decision making of the clients and other supporting stakeholders.

Scheme Design Phase

After the choice and selection of a particular design at the outline proposal stage, the scheme design stage further develops the idea of the client, including the preliminary designs from the specialist. Full, reliable and realistic cost plans and estimates of various elements, components, and parts of the project are prepared at this stage, based on the availability of the necessary information. The following are expected to be available at the completion of scheme design stage:

- Detailed specifications of various elements, sections and parts of the building;
- Visual realization of the project with detailed plans, elevations, sections, and views; and
- Elemental breakdown of project cost, known as the cost plan.

Detailed Design Phase

This is the final stage of design where the complete design for the project is prepared and produced for final decisions and actions by the client and other stakeholders. Depending on the nature of procurement and type of project, engineers, contractors, and other design specialists are also involved to complement the original design of the project. Complete and detailed cost checking is also carried out at this stage and it should be noted that any change or modifications after this phase may constitute a threat to the project in the area of re-work, cost overrun, and time overrun.

According to Kam Shadan (2012), the final design should ensure the accomplishment of the following key objectives:

- Design is biddable, constructible, and cost-effective;
- Designers' feedback before progressing further;
- Coordination between engineering disciplines;
- Building codes and regulations compliance;
- Adherence of cost estimates to the budget;
- Operational and functional objectives met;
- Identification of errors and omissions;
- Quality of the design;
- Adherence to design criteria and environmental documents;
- Interface compatibility in respect to adjacent project elements and the existing transit system; and
- Compliance of final construction contract documents with standards including design criteria, environmental document, codes, and regulations.

Production Information Phase

Depending on the procurement method adopted for the project, this stage entails the preparation of all documents and requirements that are needed for the invitation of tenders from contractors. The production information comprises documents relating to the final details of the project which include working drawings, specifications from various consultants, schedules of work, conditions of contract, and a bill of quantities. According to the Treasury Government Office (2002), the final design drawing may include the following:

- Site plan;
- Location plan;
- General arrangement plan;
- Foundation layout;
- Detailed floor plan;
- Detailed sections;

- Plan of external landscaping;
- Plan of key building function;
- Detailed elevations;
- Frame details;
- Drainage details;
- Key details drawings;
- Details of furniture and fittings; and
- Assembly and component drawing.

Moreover, the final specification information should include the following:

- Specification for main elements or components;
- Schedule of finishes;
- Statement of required quality;
- Acoustic or vibration requirements;
- Details of alternative specifications;
- Final performance criteria for main elements or components;
- Final specification for services;
- Structural performance criteria;
- Specification for plants and equipments; and
- Final specification for materials and finishes.

Other documents at this stage include the following:

- Bills of quantities for main contract and service;
- Standard form of tender;
- Form of agreement; and
- Condition of contract.

Tender Action Phase

The purpose of tendering is to obtain quotations regarding project cost, time, expertise, and other required details from the contractors in respect of their ability to execute the project based on the available production information documents. The tendering method can be open or closed, competitive or non-competitive, selective or non-selective, and single or multiple stage, depending on factors relating to the project and choice of the client. There are the following three stages in tendering for construction projects:

- Invitation to offer;
- Offer; and
- Acceptance of offer.

Using any of the tendering methods highlighted earlier, the first stage of tender action is the invitation of the contractors, also known as the tenderers, to offer (bid) for the job through an advertisement in newspapers, magazines, an international website, or the like. The essence of the production information at this stage is to ensure that all the contractors tender and price on the same basis without bias and constraints.

The second and most important phase to the contractor is the decision to offer for the job. Factors which are considered at this point include those highlighted in the 'objectives of construction projects' section of this chapter. However, the decision to tender or not depends on the management members of the contracting firms and is determined by their experience, technical expertise, and the expected and foreseen profit, among other factors.

Based on specified and standard criteria relevant to the project, it is the responsibility of the client and his or her representatives to select an appropriate contractor who will be fitting for the project. Such factors considered in the acceptance of the contractor's offer may include technical expertise, experience of the area, previous experience of similar projects, ability to command or provide funds for the project, and sustainable construction experience. At the selection of successful contractors, the following information is also required for the firms and reference for future actions: the names of contractors involved in the tendering process; the name of the successful bidder; and the selection approach adopted.

Project Planning Phase

One of the required documents expected to be submitted by the contractor during the tender action phase is the plan of construction activities, also known as the programme of works. At the point of acceptance, this becomes a major contract document that is expected to guide the activities of the construction project as agreed by the contractor, client, and other participants.

At this stage, the contractor can also organise a pre-construction conference attended by representatives of the client and other principal participants who influence or will be affected by the project (The Associated General Contractors of America 2003). The purpose of the meeting includes the following:

- Introducing members to one another for cordial communication and promoting effective relationship for project success;
- Reviewing programme of works or job schedule for review, corrections, adjustments;
- Coordinating submission procedure of drawings and other necessary documents during the construction process;
- Establishing procedures for change order implementation;
- Raising queries that require attention by participants for the smooth running of the construction activities;
- Identifying and evaluating likely problems with labour and materials and suggest solutions accordingly;

- Establishing and agreeing on how non-contracted services will be executed and paid for. These includes temporary facilities, clean-ups, etc.;
- Identifying project temporary facilities and procedure for their usage and maintenance; and
- Explaining project list procedures.

It is essential for the planning meeting to be appropriately and adequately planned with the inclusion of relevant stakeholders. If the preconstruction meeting is well planned and executed, it will help to:

- Establish better lines of communication;
- Establish clear understanding of requirements;
- Unify management requirements;
- Avoid disputes and conflicts that may arise as a result of unsettled issues;
- Avoid or eliminate delay and disagreement that may arise in the course of executing the project; and
- Agree on means to ensure project are delivered to time, cost, quality and other performance measures.

Site Operation Phase

The operation on site is the actual construction phase where the expertise of the contractors and other participants is amassed and harnessed for the sole purpose of achieving the objective of the project. Depending on the arrangement, works executed are valued and paid at different stages while the actual rate and performance of works are compared with the planned activities.

Completion Phase

The completion of the project may take different forms. For instance, depending on the signed contract, the completion of a part of a project such as a 'gate house' or 'building first floor' may signify a completion phase and this is known as partial completion. The practical completion is when all the agreed activities with respect to the project, including various forms of re-work, variation, and additional works, are deemed to be satisfactorily completed. A certificate of practical completion is expected to be offered to the contractor, usually after the defect liability period.

Handover and Feedback

Some authors believe that the construction process ends at the completion phase but the issues of handling over, feedback, operation, re-use, and demolition have become major stages in the planning and execution of construction projects. Handover is the actual process of delivery project to the client or whoever commissioned it and it is the transition between the completion and operation phase (Kandeil et al. 2010; Queensland Department of Public Works 2010; Kam Shadan 2012; Oke et al. 2016). Handling over and feedback activities include the following:

- Commissioning;
- Receipt of building documentation;
- Project audit;
- Production of as-built and installed drawing;
- Operation and maintenance manual;
- Health and safety file;
- Plant, equipment and service schedule;
- System information schedule;
- Final inspection;
- Whole-life costing information;
- Service maintenance requirements;
- Certificates, warranties and guarantees;
- Occupational instructions;
- Training packages and manual;
- Final payment; and
- Resolve outstanding change and claim disputes.

Feedback is also known as closeout. At this stage, all the participants to the project are expected to submit reports that will be incorporated into the final report by the lead or prime consultant. The objective of the handover and feedback is to ensure that project activities are completed to cost, time, quality, and satisfactory standard as well as to gather necessary and relevant information on various stages of the project for the purpose of evaluating them to improve performance of future similar projects (Treasury Government Office 2002). At the handing over and feedback stages, it is possible that some omissions and defects are noticed and notified, especially during the liability period. It is the responsibility of the contractors to ensure that such issues are addressed while the clients' representatives monitor and ensure compliance to standard.

Operation and Usage Phase

After taking possession or ownership of the project, the next phase of a project is the usage and operation which involves putting the project to its original and intended use as planned during the preconstruction phase. Some of the documents highlighted at the feedback and handling over are very important at this stage to guide the activities to which the project will be subjected. There is also a need for various forms of maintenance, including planned and unplanned, corrective and preventive, to be discussed at this phase of the project life cycle.

Re-use and Demolition

Another stage of construction process is the re-use to which the project may be subjected after completion of the initial or original purpose. These may be due to an unforeseen weakness of the structure, market conditions prevalent at the time, or changing levels of demand by customers or users. However, the re-use should be in accordance with standard practice and regulations. The use may also be related to the re-use of various parts of the projects after demolition. This is fundamental to the sustainability principle and should be factored in at the project planning stages to ensure that the parts or components of the project do not only serve the purpose of the project and then become a burden to the environment afterwards, but they should also be useful for further purpose or recycling.

Attributes of Construction Projects

The basic characteristics of construction projects are highlighted and described in this section based on the existing studies by various authors (Chan 2001; Takim and Akintoye 2002; Leung et al. 2010; Kam Shadan 2012). This is as a result of the involvement of numerous and diverse parties, interwoven phases and stages, systematic and logical processes as well as the involvement and participation of the public and private sector of the economy. The attributes include being dynamic, multidisciplinary, multifaceted, schedule based, risky, waste generating, expensive, complex, unpredictable, and stressful.

Dynamic

The changing nature of construction projects is one of the major attributes that should be considered, especially when there is a need to compare one project with another. No two projects are the same, even if they are of the same design, belong to same client with same consultants and located in the same area. The dynamism of construction projects is as a result of increasing uncertainties in the development process, budgets and technology.

Scope

This is related to the specific need the project is to meet, coupled with the level of service and regulatory requirements expected of various projects: these make construction projects differ from one another. Moreover, each project is unique and written documents that specify regulatory requirements, operational needs, and specifications should be provided specifically for the project under consideration.

Multidisciplinary

Construction projects involve the participation of various kinds of stakeholders, including those with direct and indirect links. There are various kinds of contractors, subcontractors, consultants and regulatory bodies working together for the successful delivery of construction projects.

Multifaceted

There are various facets of a construction project, though the major interest is to achieve value for money for construction client and ensure return on investment: various stakeholders perceive the construction projects based on their own experience, training and personal interest. Moreover, there are various features, parts and aspects which are to be considered in the planning, development, and delivery of construction projects.

Schedule Based

All projects have a definite start and end time: this issue is fundamental to construction projects because of the level of involvement of valuable resources. A slight change in the planned schedule of construction projects may lead to time overrun which eventually results in various challenges, including project abandonment, disputes, and conflicts.

Risky

Construction is a risky business and the risks emanate not only from the process and materials but also from stakeholders and participants to the project. Various construction management tools, such as procurement methods, tendering techniques, and the adoption of various contract forms, are means of managing the risks of construction projects

Waste Generating

One of the issues with the construction industry is the generation of high levels of both physical and non-physical waste. Over the years, there has been the introduction of regulations and policies to manage the non-physical waste through the advent of new and modern equipment. There has also been an increase in the re-use of physical waste for other construction- and non-construction-related activities.

Expensive

Most construction projects are capital intensive and are also constrained by budgetary limits by the client, sponsor or financier. One of the major considerations is to ensure value for money for construction clients by not only making sure that the project recoups its original money but that it is profitable afterwards.

Complex

Owing to the employment of various forms of plant and equipment as well as diverse individuals with differing interests, construction projects are complex in nature. Also, designs are becoming more competitive, and the introduction of new regulations relating to sustainability and other new measures have increased the complex nature of construction projects.

Unpredictable

It is necessary to plan exclusively for construction projects and various methods have been introduced to achieve this in the industry. However, despite the use of feasibility studies and investment appraisal techniques, some aspect of construction projects are still expressed as provisional, especially the substructure aspect. This is evident in the nature of variation and additional works associated with construction works generally in some economies, if not all.

Stressful

The execution of construction projects has been described as stressful (Leung et al. 2010) as a result of poor working environments of construction projects, complicated teamwork relationships among different parties, complexity of tasks, and increasing level of targets due to changing client demands, among others.

Stakeholders in Project Construction

Stakeholders in the construction industry are participants or individuals involved in and affected by the activities of construction projects throughout the whole-life cycle of the project, that is, from preconstruction, construction, and post-construction, for the purpose and during the course of actualising the objective of the project. There are primary, key supporting, tertiary and extended team members in order of priority to a construction project. Nash et al. (2010) classify them as internal (demand and supply side) and external (private and public) stakeholders. Also, based on attributes of urgency (urgent claim), legitimacy (legitimate relationship) and power (ability to impose), there are eight classes of stakeholders (Mitchell et al. 1997; Olander and Landin 2005; Malkat and Byung-GYOO 2012; Aapaoja and Haapasalo 2014). They include the following:

- Stakeholders that have neither urgency, legitimacy or power, and therefore not really stakeholders;
- Demanding stakeholders with urgent claim but no power and legitimacy;
- Discretionary stakeholders have the legitimacy attribute but lack power and urgency;
- Dormant stakeholders have power but no legitimate relationship and urgency;
- Dependent stakeholders have the urgency and legitimacy but no power;
- Dominant stakeholders are powerful and legitimate but lack urgent claim;
- Dangerous stakeholders lack legitimacy but possess power and urgency; and
- Definitive stakeholders possess the three attributes.

The construction industry is complex and an all-comers affair involving clients, owners, sponsors, contractors, construction professionals, businessmen, the community, and regulatory bodies. This necessitates the delegation of duties and responsibilities to individuals as persons or organisations (Gluch 2009;

Bal et al. 2013). Authors of construction-related studies have identified various stakeholders with direct and indirect links to project construction (Treasury Government Office 2002; Nash et al. 2010; Nash et al. 2010; Oke et al. 2010; Hewage et al. 2011; Kam Shadan 2012; Zanjirchi and Moradi 2012; Bal et al. 2013; Oke 2013; Oke et al. 2013; Nyandika and Ngugi 2014; Olatunji et al. 2014a).

There is complexity in the engagement and management of construction stakeholders due to the potential and expected number of stakeholders of a typical construction project. They are numerous, diverse, multidisciplinary and disparate. Typical stakeholders in the construction industry include the following, among others: client, owner, sponsor, financier, principal contractor, trade contractor or subcontractor, material supplier, subcontractor, architect, quantity surveyor, employee, engineer, archaeologist, sustainability consultant, development manager, local government, national government, design coordinator, regulatory agency, managing director, technical director, conservationist, environmentalist, project manager, area manager, builder, construction manager, project manager, land surveyor, estate surveyor and other specialist consultants. Others are client's customers, client's employees, client's tenants, client's suppliers, local residents, and local landowners among others.

For construction projects to be successful, the stakeholders – especially the direct participants – are expected to be proactive in discharging their responsibilities. The major influencing factors of their performance include the nature and type of project, the type of client, and perceptions of other participants. Apart from characteristics relating to project and client, there are other basic factors that may influence the performance of construction professionals and indirectly affect the performance of the project under consideration. These are classified into two categories, namely, organisational and demographic (Olatunji et al. 2014b). The organisational determinants include remuneration, promotion opportunities, motivation, attitude of co-workers, supervision, incentives, and a sense of belonging, while the demographic descriptors are concerned with age, gender, marital status, educational level, length of service, and experience, among others. The project performance of contractors can be determined using integration performance and specification performance; human performance; organizing performance; and technical performance (Zanjirchi and Moradi 2012).

Team Roles of Construction Project Stakeholders

One of the key issues with stakeholders' management is their ability to work and operate as a team. Right from the ancient times, humans have worked together in a team to achieve most of their set goals and objectives to their satisfaction. As explained in chapter three, having the right mix of team roles is an important factor for a successful value management study. The principle of teamwork is also fundamental to the success of construction projects as it involves people of various disciplines working together to achieve the objective of projects, centred on the provision of value for money for construction clients. However, depending on factors such as trust, conflict, commitment, accountability, and togetherness, a team can be considered healthy or dysfunctional, that is effective or ineffective (Belbin 1981; Lencioni 2005; Oke and Ukaeke 2013). This underscores the need to explain various team roles expected of team members, be it a construction or value management team.

The success of a value management exercise largely depends on the understanding, willingness, and readiness of project or design team members to incorporate various recommendations from the study into the project. This will not only be affected by the professional training of the members but also by their team roles. The team roles, according to Aritzeta et al. (2007), arose from the need for different roles to dominate at different stages of team development. The identified six stages required for team development are the following:

- Need identification;
- Ideas finding;
- Plans formulation;
- Ideas making and generation;
- Establishment of team organisation; and
- Following through.

Various authors have examined the issue of team roles in general business as well as specifically in relation to construction activities, based on the earlier work by Belbin. (Belbin 1981, 1993; Baker and Salas 1997; Cornick and Mather 1999; Constructing Excellence 2004; Carson and Isaac 2005; Ochieng and Price 2009; Olatunji et al. 2014a, b; Senaratnea and Gunawardane 2015; Oke et al. 2016b). Based on these authors' contributions as well as on the earlier work by Belbin on the behavioural pattern of individuals in a teamwork, the Belbin Team Role Self-Perception Inventory (BTRSPI) has become a major basis for measuring behavioural characteristics of individuals in teamwork. There are nine team roles, namely plant, specialist, completer or finisher, implementer, team worker, monitor or evaluator, shaper, coordinator, and resource investigator.

The roles are classified in various ways. One way is the use of primary and secondary distinctions while another is preferred, manageable and least preferred, based on the disposition and attributes of individuals. They are also classified as action, social, and thinking roles, based on the expectation required of the individuals who possess the attributes. Action roles comprise completer or finisher, shaper, and implementer while social ones are related to coordinator, team worker, and resource investigator. The thinking group comprises the remaining three roles, namely, specialist, plant, and monitor evaluator.

Plant

Plant, also referred to as planter by some scholars, is an important role required for a team to be effective, efficient and successful. 'Plants' are more concerned with issues and actualising them than the details surrounding them and possess fast thinking ability to solve problems. The basic attributes of a plant include the following:

- Imaginative;
- Creative;
- Introvert;
- Original;
- Radical-minded;
- Uninhibited;
- Dominant; and
- Trustful.

Specialist

Specialists are known to have a peculiar interest in a particular area and are often less useful in others. They are master of their specialty and possess an in-depth knowledge of their area with a high level of competence and experience. A specialist possesses the following attributes:

- Expert;
- Efficient;
- Focus;
- Defendant;
- Serious; and
- Dedicated.

Completer or Finisher

This is an attribute of an individual who is only interested in making sure that ideas and plans are executed and concluded in the right manner while following the acceptable standard. Completers are more interested in accuracy than precision. The basic attributes of a completer or finisher include the following:

- Conscientious;
- Submissive;
- Anxious;

- Self-disciplined;
- Self-controlled; and
- Perfectionist.

Implementer

The implementer team member is also known as company worker. The implementer is more interested in ensuring that ideas and concepts from the activities of the team are executed without delay. They usually take up activities that do not interest others. An implementer is expected to have the following attributes:

- Conservative;
- Systematic;
- Sincere;
- Inflexible;
- Logical;
- Reliable;
- Assertive;
- Realistic;
- Disciplined;
- Efficient; and
- Sincere.

Team Worker

A team worker is any individual who is interested in the cohesion and smooth running of the team, even if it will be at his or her expense. They are needed to keep the team together by resolving conflicts and disputes while using a people-oriented approach and diplomatic skills to keep members of the team working as one. A team worker is required to possess the following characteristics:

- Likeable;
- Uncompetitive;
- Unassertive;
- Listener;
- Loyal;
- Supportive;
- Teachable;
- Submissive;
- Accommodating; and
- Diplomatic.

Monitor or Evaluator

A monitor pays attention to detail and evaluates the activities of the group and individuals to ensure that the right decisions are taken. The skills of monitors or evaluators are better felt during a crucial decision-making stage where their ability to examine and evaluate related and competing ideas becomes crucial to the team. The traits of a monitor or evaluator include the following:

- Reliable;
- Prudent;
- Dependable;
- Unambitious;
- Trustworthy;
- Open-minded;
- Judicious;
- Fair-minded;
- Serious; and
- Intelligent.

Shaper

A shaper has a high motivation for focusing on a task and uses various means such as dialogue, brainstorming, and argument to ensure that the task is well delivered. A shaper possesses the following traits:

- Bold;
- Abrasive;
- Edgy;
- Impulsive;
- Arrogant;
- Dominant;
- Competitive;
- Emotional;
- Anxious;
- Self-confident; and
- Impatient.

Coordinator

The coordinator is also known as chairman. A coordinator is a manager, planner, controller and arranger of not only individuals who made up the team but also the activities of the team, including ideas and concepts. A coordinator is usually the leader of the team and, using various available leadership styles, he or she brings together members of the team and ensures that everyone is working on a task for the achievement of team goals and objectives. Owing to the leadership role, the following attributes are expected of a coordinator:

- Trusting;
- Extrovert;
- Positive thinker;
- Confident;
- Committed;
- Stable;
- Positive;
- Dominant;
- Self-disciplined;
- Mature; and
- Assertive.

Resource Investigator

A resource investigator considers and explores different means of getting information and opportunities to support and actualise ideas and concepts emanating from the team. A resource investigator makes a lot of contact, undertakes research, carries out market surveys and gets involved with other activities geared at discovering and exploring resources outside the team but which enhance their performance and make the team successful by achieving their goals and objectives. A typical resource investigator is expected to possess the following attributes:

- Diplomatic;
- Optimistic;
- Dominant;
- Positive;
- Stable;
- Inquisitive;
- Enthusiastic;
- Extrovert;
- Flexible; and
- Persuasive.

The importance of team roles to the effective management of construction projects cannot be over-emphasized. According to Aroba and Wedgewood-Oppenheim (1994), their importance includes the following:

- Team roles of project team members aid coordination and enhancement of the corporate standard of the construction industry to bring it at par with other corporate ones such as banking, manufacturing, and the like.
- Knowledge, understanding and awareness of team roles by members of project team allow the team members to adopt any of the preferred or manageable team roles.
- The ability to adapt and work within preferred team roles improves the productivity of construction participants and subsequently improves overall project performance.
- Understanding and adoption of team roles by construction team members enhance effective communication and good relationships.
- Improvement in primary or professional as well as in team roles by team members through their awareness and willingness to adopt their preferred roles.
- Commitment from team members exhibiting their team roles adequately is essential for the performance of the project.

An awareness, understanding and adoption of team roles also pose some challenges to the team and construction project at large. Such challenges include the following:

- The selection and composition of the project team at the initial stage do not always consider team roles of members. This may result in the absence of one or more of the required roles which may affect the team productivity and subsequently influence the performance of the project under consideration.
- A clash of interest among team members acting as coordinator and approved leader of the team, which may be the prime consultant, project manager, architect, construction manager or any other.
- The presence of more members of the team performing the same team role poses a danger to the team, especially if some other important roles are absent.
- Owing to the rigorous pursuit of the team's objective, conflict, disputes and aggravation may arise if there are two or more shapers in the team.
- It is possible for team members to exhibit more than one team role and find it challenging to specialize in a particular one.
- Team roles such as plant and coordinator are domineering and members exhibiting these roles may be envied by others, resulting in team dysfunction, ineffectiveness, and low productivity and performance.
- Over-commitment to a team role may be an advantage to the team objective since the teamwork goes beyond the actualisation of just one of the team roles but joint effort is required to achieve the project objective.

Summary

Owing to new and emerging performance measurements for construction projects, clients of projects related to the construction industry will continue to demand more complex projects and it is the responsibility of the participants, especially the contractors and consultants, to ensure that the industry continues to be relevant in the face of changing demands, needs and requirements. One of the means of achieving this is through adhering to construction processes and ensuring that the right mix of team members in terms of team roles and professional training is in place for every construction work.

This chapter explained the concept of construction projects holistically and therefore provides a background for chapter five on measures of project success as well as chapter six which explains the concept of sustainability in construction. This is geared towards the discussion provided in chapter eight explaining how value management can be adopted for the sustainability of construction projects.

References

Aapaoja, A., & Haapasalo, H. (2014). A framework for stakeholder identification and classification in construction projects. *Open Journal of Business and Management, 2*(2014), 43–55.

Aritzeta, A., Swailes, S., & Senior, B. (2007). Belbin's team role model: Development, validity and application for team building. *Journal of Management Studies, 2*(2), 2362–2380.

Aroba, A., & Wedgewood-Oppenheim, O. (1994). Construction team management. *Journal of Quarterly Construction Report, 2*(1994), 15–20.

Baker, D. P., & Salas, E. (1997). *Principles for measuring teamwork: A summary and look toward the future*. Mahwah, NJ.: Lawrence Earlbaum Associates Inc.

Bal, M., Bryde, D., Fearon, D., & Ochieng, E. (2013). Stakeholder engagement: Achieving sustainability in the construction sector. *Sustainability, 6*(5), 695–710.

Banaitiene, N. & Banaitis, A. (2012). Risk management in construction projects. In N. Banaitiene (Ed.), *Risk management—current issues and challenges* (pp. 429–448). London: Intech Open Science. Retrieved September 14, 2016, from http://cdn.intechopen.com/pdfs-wm/38973.pdf.

Belbin, R. M. (1981). *Management teams: Why they succeed or fail*. London: Heinemann.

Belbin, R. M. (1993). *Team roles at work*. Oxford: Butterworth-Heinemann.

Carson, A. & Isaac, M. (2005). *A guide to team roles: How to increase personal and team effectiveness*. Retrieved June 4, 2016, from http://resources.3circlepartners.com/h/i/110944715-a-guide-to-team-roles.

Chan, A. P. (2001). *Framework for measuring success of construction projects*. Brisbane. Retrieved November 27, 2016, from www.construction-innovation.info.

Excellence, Constructing. (2004). *Effective teamwork: A best practice guide*. London: Constructing Excellence.

Cornick, T., & Mather, J. (1999). *Construction project teams: Making them work profitably*. London: Thomas Telford.

Gluch, P. (2009). Unfolding roles and identities of professionals in construction projects: Exploring the informality of practices. *Construction Management and Economics, October* (2009), 959–968.

Hewage, K. N., Gannoruwa, A., & Ruwanpura, J. Y. (2011). Current status of factors leading to team performance of on-site construction professionals in Alberta building construction projects. *Canadian Journal of Civil Engineering, 38*(2011), 679–689.

Isle of Man. Treasury Government Office. (2002). *Procedure notes for management of construction project.* Douglas: Treasury Government Office. Retrieved September 14, 2016, from https://www.gov.im/media/383284/procedure_notes_for_management_of_construction_projects.

Kam Shadan, P. E. (2012). *Construction project management handbook.* Washington: Federal Transit Administration. Retrieved September 14, 2016, from https://www.transit.dot.gov/sites/fta.dot.gov/files/FTA_Report_No._0015_0.pdf

Kandeil, R., Hassan, M. K. & Nady, A. E. (2010). Hand-over process improvement in large construction projects. *Facing the challenges, building the capacity* (pp. 18–28). Sydney: International Federation of Surveyors (FIG) Congress.

Lencioni, P. (2005). *The five dysfunctions of a team.* San-Francisco: Jossey-Bass.

Leung, M.-Y., Chan, Y.-S., & Chong, A. M. (2010). Chinese values and stressors of construction professionals in Hong Kong. *Journal of Construction Engineering and Management, 136*(12), 1289–1298.

Malkat, M. & Byung-GYOO, K. (2012). *An investigation on the stakeholders of construction projects in Dubai and adjacent regions.* Retrieved September 14, 2016, from http://www.ipedr.com/vol45/016-ICMTS2012-M00008.pdf.

Mitchell, R. K., Agle, B. R., & Wood, D. J. (1997). Towards a theory of stakeholder identification and salience: Defining the principle of who and what really counts. *The Academy of Management Review, 22*(4), 853–886.

Nash, S., Chinyio, E., Gameson, R., & Suresh, S. (2010). The dynamism of stakeholders' power in construction projects. In C. Egbu (Ed.), *26th Annual ARCOM Conference* (pp. 471–480). Leeds: Association of Researchers in Construction Management.

Nyandika, F. O., & Ngugi, K. (2014). Influence of stakeholders' participation on performance of road projects at Kenya National Highways Authority. *European Journal of Business Management, 1*(11), 384–404.

Ochieng, E. G., & Price, A. D. (2009). Framework for managing multicultural project teams. *Engineering Construction and Architectural Management, 16*(6), 527–543.

Oke, A. E. (2013). Project management leadership styles of Nigerian construction professionals. *International Journal of Construction Project Management, 5*(2), 159–169.

Oke, A. E., & Ukaeke, I. L. (2013). Factors responsible for effective and ineffective teams in Nigerian construction industry. *Journal of Construction Management, 28*(4), 5–16.

Oke, A. E., Ibirobke, O. T., & Aje, I. O. (2010). Perception of construction professionals to the competencies of Nigerian quantity surveyors. *Journal of Building Performance, 1*(1), 64–72.

Oke, A. E., Ogunsemi, D. R., & Adeeko, O. C. (2013). Assessment of knowledge management practices among construction professionals in Nigeria. *International Journal of Construction Engineering and Management, 2*(3), 85–92.

Oke, A. E., Olatunji, S. O. & Ajulo, A. A. (2016a). Factors affecting construction project handover and feedback mechanism. In O. J. Ebohon, D. A. Ayeni, C. O. Egbu & F. K. Omole (Eds.), *21st Century Human Habitat: Issues, Sustainability and Development. Proceedings of Joint International Conference* (pp. 842–850). Akure.

Oke, A. E., Olatunji, S. O., Awodele, A. O., Akinola, J. A. & Kuma-Agbenyo, M. (2016b). Importance of team roles composition to success of construction projects. *International Journal of Construction Project Management, 8*(2), 141–152.

Olander, S., & Landin, A. (2005). Evaluation of stakeholder influence in the implementation of construction projects. *International Journal of Project Management, 23*(4), 321–328.

Olatunji, S. O., Akinola, J. A., Oke, A. E. & Osakuade, A. O. (2014a). Construction professionals' team roles and their performance. *International Journal of Advanced Technology in Engineering and Science, 2*(8), 308–316.

Olatunji, S. O., Oke, A. E. & Owoeye, L. C. (2014). Factors affecting performance of construction professionals in Nigeria. *International Journal of Engineering and Advanced Technology, 3*(6), 76–84.

Queensland Department of Public Works. (2010). *Handover: Guidance for commissioning and handover associated with government building projects*. Queensland: Queensland Department of Public Works.

Senaratnea, S., & Gunawardane, S. (2015). Application of team role theory to construction design team. *Architectural Engineering and Design Management, 11*(1), 1–20.

Takim, R., & Akintoye, A. (2002). Performance indicators for successful construction project performance. In D. Greenwood (Ed.), *18th Annual ARCOM Conference* (pp. 545–555). Northumbria: Association of Researchers in Construction Management.

The Associated General Contractors of America. (2003). *Guidelines for a successful construction project.* Washington: The Associated General Contractors of America, the American Subcontractors Association, and the Associated Specialty Contractors. Retrieved September 14, 2016, from https://www.discountpdh.com/course/guideline-on-general-contractor-subcontractorrelations.

United Kingdom. Cabinet Office. (2011). *Government construction strategy.* London: Cabinet Office. Retrieved November 14, 2016, from http://www.gov.uk.

United Kingdom. HM Treasury. (2014). *Infrastructure cost review: Measuring and improving delivery.* London: HM Treasury. Retrieved September 14, 2016, from http://www.gov.uk.

Zanjirchi, S. M., & Moradi, M. (2012). Construction project success analysis from stakeholders' theory perspective. *African Journal of Business Management, 6*(15), 5218–5225.

Chapter 5
Measures of Project Success

Abstract Every project is initially conceived and designed to achieve a specific objective relating to the desire and need of the client. However, owing to various interests of other construction stakeholders, the objectives of the projects become numerous, depending on the views of the participants. These objectives are referred to as measures, factors, methods, criteria, indices, or attributes of project success, performance, or delivery. Generally, the success of construction projects is influenced by various factors. They are classified as project characteristics, contractual arrangements, project participants, or interactive processes. This chapter therefore highlights and explains various measures of project success with a conclusion that sustainability goals encompass every other measure.

Keywords Project cost · Project duration · Project objective · Project quality · Project success · Sustainable goals

Introduction

Every project is initially conceived and designed to achieve a specific objective relating to the desire and need of the client. However, owing to various interests of other parties as discussed in Chap. 4, the objectives of the projects become numerous, depending on the views of the participants. These objectives are referred to as measures, factors, methods, criteria, indices, or attributes of project success, performance, or delivery. In this book, 'measures of project success' is adopted to give a clear meaning and differentiate the measures from project delivery methods identified as design-bid-build (DBB), construction manager/general contractor (CM/GC) and design-build (DB). To stress the important ones, some construction management scholars have adopted the terms 'key' and 'critical' in describing the measures such as critical success factors (CSFs), key performance indicators (KPIs), and the like. However, it is believed that all measures are important, relevant, crucial, critical, and vital, depending on the type of projects and their objectives.

A.E. Oke and C.O. Aigbavboa, *Sustainable Value Management for Construction Projects*, DOI 10.1007/978-3-319-54151-8_5

The measures of project success is a term adopted for the objective of all roles, responsibilities and contractual relations of the entities involved in a project (Touran et al. 2009). The Associated General Contractors of America (2003) explained project success measures as the complete and all-inclusive performance measurement process of delegating contractual responsibilities and duties for designing and construction of projects. This implies that project success measures are functions of technical relationships in respect of a contract agreement among project stakeholders, especially clients, owners, sponsors, financiers, statutory bodies, consultants, and contractors.

Success of Construction Project

There are individual and collective goals for everyone involved in a project. For construction projects, the measure of their success can be viewed from two angles. The first is the view of participants who have a vested interest in and direct impact on the activities involved in the project. These individuals or organisations can control and direct the activities in their favour, depending on their interest and level of influence. Apart from overall project success objectives, each of these individuals or groups of people also measure the success of the project by virtue of the achievement of their individual or organisational goals. However, there are other groups of stakeholders who are affected by the project, either at the preconstruction, construction or post-construction stage, but have little or no direct influence on the process itself. From these explanations, what constitutes construction project success varies from one group of individuals to another. In view of this, Tabish and Jha (2011) note that achieving success, especially for public construction projects, is more difficult and demanding owing to the level of transparency, fairness, efficiency, economy, and quality required.

The traditional measures of construction project success are still the most popular, they are based on the iron triangle principle that include cost, time, and quality (De Wit 1988; Atkinson 1999; Aibinu and Jagboro 2002; Yu eta al. 2005; Toor and Ogunlana 2008; Ahsan and Gunawan 2009; Olawale and Sun 2010; Tatum 2011; Doloi 2012; Kog and Loh 2012; Oke 2016). However, the increasing number of performance measures demanded by clients and other stakeholders has led to the emergence of several others. In fact, the iron triangle measures have been described as quantitative variables which makes them inadequate and narrow because they fail to capture other aspects relating to subjective criteria based on the opinions of stakeholders, the macroeconomic benefits of reliability, and functionality, and other non-quantitative criteria (Atkinson 1999; Lim and Mohamed 1999; Miller et al. 2000; Chan 2001; Shenhar et al. 2001; Takim and Akintoye 2002; Dainty et al. 2003; Chan and Chan 2004; Diallo and Thuillier 2004; Hughes et al. 2004; Ahadzie 2008; Doloi 2009; Hale et al. 2009; Kim et al. 2009; Chen et al. 2010; Tabish and Jha 2011; Thomson 2011; Toor and Ogunlala 2010; Chen et al. 2011; Tenah 2011;

Hwang and Lim 2012; Kog and Loh 2012; Ren et al. 2012; Shahu et al. 2012; Zanjirchi and Moradi 2012; Davis 2013; Oke et al. 2016).

The measures are classified as objective and subjective, micro and macro, delivery and post-delivery, qualitative and quantitative, technical, performance and overall, perception and reality, among others. According to scholars, experts and researchers in the construction industry, the measures of project success include cost, operational performance, time, quality, satisfaction, as well as health and safety among others.

From these measures, various variations have also emerged: these include value, profit, speed of construction, accident rate, net present value, flexibility, free from defect, fitness for purpose, social obligation, contractors' performance, percentage net variation over cost, time variation, unit cost, absence from legal claims and proceedings, environmental impact assessment (EIA) scores, functionality, research and development, training and recruitment, construction team's satisfaction, design team's satisfaction, client's satisfaction with product, client's satisfaction with services, and end-user's satisfaction. Moreover, over the years new criteria have been introduced to guide the performance of construction projects. These include sustainability, energy saving/efficiency, and carbon reduction.

A further examination of the highlighted measures reveals that most of them are inter-dependent in that they affect one another. For instance, a construction project that experiences time overrun is most likely to also experience increase in cost, especially in the area of preliminaries. Also, measures such as satisfaction depend on such variables as cost, time, quality, as well as health and safety. However, the basic ones of cost, quality, time, operational performance, value and profit, health and safety, satisfaction, as well as sustainability are discussed below.

Cost

Cost is one of the fundamental bases for determining the success of construction projects. The cost of a construction project is usually established at the preconstruction stage and may depend on the budgetary limit of the client. Alternatively, it can be established through the application of various cost-estimating principles by a construction cost estimator. However, the most important factor is that target price or cost estimates are set and agreed upon before the planning and actual construction phase. It is believed that the cost should be a whole-life costing encompassing all related costs of initial, running, maintenance, and demolition, as well as the re-use of all the elements, components, and parts of the project. The initial cost comprises the cost of site; the cost of actual work; legal, agency and statutory fees; the cost of services also known as non-incentive based compensation; and appropriate contingencies.

Using the target price or budgetary limit, it is expected that a cost plan is prepared using an elemental or comparative technique. The former is also known as 'designing to a cost' while the latter is referred to as 'costing a design'.

The elemental cost plan entails setting a target price for the construction project which must not be exceeded in the course of actual construction. The target price, which is also known as the cost limit or estimate, can be derived at using various approximate single or dual estimating techniques such unit, and cube methods. It can also be set by the client or through a sharing formula in the case of multiple projects competing for the same capital. The principle is that every construction project consists of various elements and cost is associated with each of the elements, based on a cost analysis of similar existing projects. This is usually expressed in cost per area of gross floor area of the project and the idea is that summation of the elemental costs should not exceed the target cost.

For comparative cost planning, the emphasis, unlike the elemental method, is not really on target costs but on investigations of various parts of design and the presentation of alternatives for the satisfaction of the client's requirements in terms of specification, construction, and function. The designer is concerned with the production of an optimum design and as a result, the cost limit is usually treated as an upper limit of the construction project. The estimate is also prepared using the same principles discussed under the elemental method. Regardless of the adopted method, it could be noted that there are three phases that are considered in cost planning and cost control of construction projects. These are the following:

- Establishing the project target cost;
- Preparing the cost plan; and
- Checking the cost.

Before the final production information of the project, various adjustments are made to the derived cost plan to ensure that the project cost is within the target cost. Also, during the course of the construction, cost control and checking are usually carried out to compare the planned and actual spending with the sole aim of ensuring that the budget is not exceeded. The techniques adopted for this exercise include unit costing, actual labour, plan and material versus planned reconciliation, profit or loss on each contract at valuation dates, overall profit or loss, project cost-value reconciliation, leading parameter method, the programme evaluation and review technique (PERT), earned value analysis, and standard costing (Olawale and Sun 2010).

At the end of the project, especially at the practical completion stage, the final account is prepared detailing various adjustments to the initial cost with respect to variations, fluctuation, changes, modifications and other additional works that would have arisen during the process of executing the project. By incorporating these items and others such as legal fees and dispute resolution fees, the objective is to arrive at the final cost of the construction project which is usually not the same as the tender or contract sum.

Using cost as the basis, a project is adjudged to be successful if the final cost of construction is within the budgeted or targeted cost. This is not in respect of other variables of project success but the focus is only on the cost. For instance, it is possible to have improved quality or function of elements during the course of construction for better and enhanced performance but such a project will still be deemed unsuccessful if the initial cost is exceeded.

Achieving construction projects for the target cost has been a major challenge to stakeholders, especially those tasked with the responsibility of managing and administering project budgets. Some of the factors responsible for this difference include the following:

- Inaccurate evaluation of project cost especially at the initial stage;
- Unforeseen circumstances such as fire incidence and other acts of God;
- Inflation of unit rates and overall contract price;
- Unstable interest rate;
- Change of scope, focus and specification of parts or whole project;
- Low level of performance productivity due to low skilled manpower or skill shortage;
- Fluctuation of currency and exchange rate;
- Legal challenges;
- Fraud and corruption;
- Design changes;
- Weak regulations and control;
- Over-specification or lack of proper understanding;
- Inefficient and bureaucratic use of competition process;
- Over-dependence on foreign resources in terms of materials, plant and human;
- Lack of clarity and direction over key decisions, especially at the preconstruction stage;
- Sudden change or discontinued flow of funds;
- Inconsistent government policy and systemic issues;
- Stop-start investment programmes;
- Conflict among project parties;
- Discrepancies in contract documentation;
- Complexity of the project;
- Lack of appropriate software to manage and control cost; and
- Emphasis on achieving project within budget rather than on focusing on value which is concerned with function at the lowest possible cost.

Judging by the above-mentioned factors, especially the last one, construction stakeholders have realised that, for construction projects, attention and focus should not just be on cost but also on value, with the main focus on achieving function at the lowest cost. This necessitates the adoption of value management to actualise the dream of not only achieving the project at the targeted cost but also ensuring value for money and an adequate return on investment.

Time

One of the items of vital construction contract information required from contractors during the bidding process is the specification of the period of time or duration it will take for the completion of a project under consideration. It is a major criterion for evaluating the contractors in the choice and selection of the appropriate one. The duration specified by the successful contractor is usually discussed and necessary adjustments are made as appropriate. With respect to the approved and agreed on duration, a programme of work is usually prepared to guide and monitor the activities of the project through rigorous and extensive planning.

Construction time refers to the number of days, weeks, months or years for the construction phase of the project, from the start of operations on site to the practical completion stage, to be completed. Like cost, the period of project activities is checked and compared (that is, planned and actual) during the project execution to ensure that the target project duration is not exceeded.

Factors that influence the duration of projects include the following:

- Unforeseen circumstances such as fire incidence and other acts of God;
- Excessive variation and fluctuation;
- Non-performance of nominated suppliers and subcontractors;
- Inaccurate estimation of project duration;
- Lack of discipline and commitment of stakeholders to schedule;
- Change of scope, focus and specification of parts or whole project;
- Fraud and corruption;
- Design changes;
- Conflict between project parties;
- Discrepancies in contract documentation;
- Complexity of the project;
- Over-dependence on foreign resources in term of materials, plant and human;
- Low level of performance productivity due skill shortage;
- Legal challenges;
- Over-specification or lack of proper understanding;
- Inefficient and bureaucratic use of competition process;
- Lack of clarity and direction over key decisions, especially at the preconstruction stage;
- Sudden change or discontinued flow of funds; and
- Stop-start investment programmes.

There are a number of measures for ensuring that construction projects are delivered to cost and time. However, they are related to observing each of the influencing factors and devising a plan to minimize or eliminate them. These measures are classified into five groups, namely predictive, preventive, corrective-preventive, corrective-predictive, and organisational (Olawale and Sun 2010).

Quality

This is one of the traditional but subjective measures of project success. In view of this, most research adopting mathematical models have adopted only the first two measures, namely, cost and time, owing to their quantitative nature. Quality in relation to construction projects can be described as a combination of attributes that are expected of the services required by participants as well as elements or components of construction projects. It is the basis for measuring fitness for purpose and guaranteeing construction projects in general to the satisfaction of clients and end-users.

Two of the important contract documents that are vital for ensuring the quality of construction projects are specification notes and contract conditions. Specifications are documents detailing the standard, scope, requirements and benchmark for various aspects, sections, trades and materials required for various elements and parts of construction works. They highlight and explain instructions concerning the construction method to be adopted, the standard of workmanship and the quality of materials to be used for construction works. It is therefore necessary to ensure that the specification document contains the necessary information peculiar to the project and it is not just a case of adopting the same document for every form of project, bearing in mind that no two projects are the same.

There are various contract conditions, including old, new and emerging ones. They include the Joint Contract Tribunal (JCT), International Federation of Consulting Engineers (FIDIC), Joint Building Contracts Committee (JBCC), General Conditions of Contract (GCC), the Engineering and Construction Contract (ECC and NEC3 packages) among others, with some specifically designed for a particular region or country while others are international. Some are also designed for a particular type of construction project such as engineering while other are designed for and are applicable to all forms, namely, building, civil, and industrial engineering works. Regardless of the type, the essence is to stipulate and guide the activities of construction stakeholders in the discharge of their duties and responsibilities to one another and the construction activities.

From the perspective of quality, a project is deemed successful provided it is completed in accordance with specifications and the adopted general conditions of contract. Quality is sometimes measured by some metrics appropriate for the type and nature of project and can only be compared to previously executed projects of a similar nature.

The following are some factors that may affect the quality of construction projects:

- Errors in design documents;
- Conflict among trades;
- Inadequacies in specification document;
- Adoption and application of wrong or old contract conditions;
- Too many amendments to standard form or condition of contract;
- Continuous changes to specification instructions;

- Lack of expertise to execute aspect of work;
- Corruption in using materials of lower standard or quantity contrary to specification instructions; and
- Strictness on delivering project to time and cost.

Operational Performance

For some construction projects, operational performance criteria are established during the preconstruction phase, especially at the outline proposal and scheme design stage although they are refined and agreed prior to the actual construction process. The operational performance measures are related to functions and are usually in term of the need of the client and the actual purpose of the construction project

Value and Profit

Value to the owner or the client of construction project can be described as the benefits that are derived from the project, mostly related to profit, which is usually the main purpose of any form of business. This relates value in this sense to profit and has been adopted by some authors and practitioners as a measure of project success. In short, a successful project in this regard is the one that is profitable to the client, depending on the period of time it takes to recoup the original capital invested, and the commencement of profit generation.

Health and Safety

Health and safety are related to accident or injuries. For construction projects, health and safety are concerned with the planning, execution, and usage of construction projects without major harm, accident, or injury to the persons involved. In view of the value placed on human lives as well as litigation and disruption that may arise as a result of accident or injury, health and safety have become a popular and well researched issue and as a result, a key measure of project success. Various organisations, agencies, regulatory boards, and other bodies concerned with the regulation and management of people and construction activities have developed various health and safety rules, guidelines, and instructions to reduce or eliminate accidents and injuries on construction sites.

Stakeholders' Satisfaction

There are various stakeholders to every construction project as discussed in Chap. 4 of this book. These include the commissioned body, participants in the actual construction, and end-users. The commissioned body are usually referred to as clients but it can also include the owner, sponsor or financier of the project and their level of satisfaction with the project in terms of meeting initial needs, profit level, quality, and time is a measure of the success of the project. For other participants such as consultants, contractors, and regulatory bodies, a construction project will be deemed successful provided it meets the target or interest of their organisation.

Every intending user of any construction project—whether working or living in it—is expected to have an expectation prior to taking possession or ownership of the project. It is necessary that these expectations are met as dissatisfied users may render the project incapable of fulfilling its original purpose and deemed a failure. The idea behind this measure is that a project that meets expectations and satisfies end-users can be termed as successful.

Sustainability

The construction industry is a major contributor to the environment. Right from the sourcing of raw materials from the environment to the physical and non-physical waste generation during and after construction, the impact of construction projects cannot be overemphasized. However, over the years, studies have shown that for better management, the issue of environment should not only be considered but a holistic view should include the social and economic dimensions. These three aspects encompass sustainable development goals.

One of the key areas for the improvement of traditional measures of success for construction projects is the adoption of sustainability goals (The American Institute of Architects 2007). Owing to the importance attached to it, sustainable design and construction are becoming mandatory for public infrastructures in some developed countries while campaigns for its adoption in other countries are gaining popularity. Sustainable criteria such as carbon footprint, alternative energies, sustainable building tool (SBTool), green globes, and leadership in energy and environmental design (LEED), can be incorporated into project objectives at the early stage of the project. The subject of sustainable construction is discussed further in Chap. 6 of this book.

Summary

Generally, the success of construction projects is influenced by various factors. They are classified as project characteristics, contractual arrangements, project participants and interactive processes. The influencing factors include the complexity of the project, level of risks and uncertainties inherent in the project, client's understanding of the needs a building can or should fulfil, participants' experience and interest, design and leadership-related factors, the level of coordination among participants, adequacy and quality of planning, appropriateness of adopted procurement, tendering, and the contractual method. The measures are therefore necessary not to address these issues but to set a target (or set of targets) that will eventually ensure that the issues identified do not affect the project.

Depending on the nature and purpose of construction projects, the choice of appropriate measure(s) to judge their success should be stipulated and understood by participants at the preconstruction stage of the project. However, a holistic and flexible approach that incorporates more than one measure is suggested with a view to improving not only the efficiency but also the effectiveness and optimisation of construction projects – both existing and newly conceptualised. In fact, agencies and parastatals of government and professional disciplines tasked with the responsibility of managing and regulating construction activities and participants have been producing various guidelines on general measures of project success with most of them emphasising the adoption of holistic measures of sustainability goals which incorporate virtually all the measures already mentioned.

This chapter therefore highlighted and explained various measures of project success and concluded that sustainability goals encompass every other measure. This therefore serves as the background for Chap. 8 where value management is discussed as a vital tool for the achievement of sustainable construction projects.

References

Ahadzie, D. K. (2008). Critical success criteria for mass house building projects in developing countries. *International Journal of Project Management, 26*(6), 675–687.

Ahsan, K., & Gunawan, I. (2009). Analysis of cost and schedule performance of international development projects. *International Journal of Project Management, 28*(1), 68–78.

Aibinu, A. A., & Jagboro, G. O. (2002). The effects of construction delays on project delivery in Nigerian construction industry. *International Journal of Project Management, 20*(2002), 593–599.

Atkinson, R. (1999). Project managemement: Cost, time and quality, two best guesses and a phenomenon, its time to accept other success criteria. *Journal of Project Management, 17*(6), 337–342.

Chan, A. P. (2001). *Framework for measuring success of construction projects.* Brisbane. Retrieved November 27, 2016, from www.construction-innovation.info.

Chan, A. P., & Chan, A. P. (2004). Key performance indicators for measuring construction project success. *Benchmarking, 11*(2), 203–221.

Chen, Y. Q., Liu, J. Y., Li, B., & Lin, B. (2011). Project delivery system selection of construction projects in China. *Expert Systems with Applications, 38*(5), 5456–5462.

Chen, Y. Q., Lu, H., Lu, W., & Zhang, N. (2010). Analysis of project delivery systems in Chinese construction industry with data envelopment analysis (DEA). *Engineering, Construction and Architectural Management, 17*(6), 598–614.

Dainty, A. R., Cheng, M. I., & Moore, D. R. (2003). Redefining performance measures for construction project managers: An empirical evaluation. *Construction Management & Economics, 21*(2), 209–218.

Davis, K. (2013). Different stakeholder groups and their perceptions of project success. *International Journal of Project Management, 31*(3), 164–170.

De Wit, A. (1988). Measurement of project success. *International Journal of Project Management, 6*(3), 164–170.

Diallo, A., & Thuillier, D. (2004). The success dimensions of international development projects: The perceptions of African project coordinators. *International Journal of Project Management, 22*(1), 19–31.

Doloi, H. (2009). Analysis of pre-qualification criteria in contractor selection and their impacts on project success. *Construction Management and Economics, 27*(2009), 1245–1263.

Doloi, H. (2012). Understanding impacts of time and cost related construction risks on operational performance of PPP projects. *International Journal of Strategic Property Management, 16*(3), 316–337.

Hale, D. R., Shrestha, P. P., Gibson, G. E., Jr., & Migliaccio, G. C. (2009). Empirical comparison of design/build and design/bid/build project delivery methods. *Journal of Construction Engineering and Management, 135*(7), 579–587.

Hughes, S. W., Tippett, D. D., & Thomas, W. K. (2004). Measuring project success in the construction industry. *Engineering Management Journal, 16*(3), 31–37.

Hwang, B. G., & Lim, E. S. (2012). Critical success factors for key project players and objectives: case study of Singapore. *Journal of Construction Engineering and Management, 139*(2), 204–215.

Kim, D. Y., Han, S. H., Kim, H., & Park, H. (2009). Structuring the prediction model of project performance for international construction projects: A comparative analysis. *Expert Systems with Applications, 36*(2), 1961–1971.

Kog, Y. C., & Loh, P. K. (2012). Critical success factors for different components of construction projects. *Journal of Construction Engineering and Management, 138*(4), 520–528.

Lim, C. S., & Mohamed, M. Z. (1999). Criteria of project success: An exploratory re-examination. *International Journal of Project Management, 17*(4), 243–248.

Miller, J. B., Garvin, M. J., Ibbs, C. W., & Mahoney, S. E. (2000). Toward a new paradigm: Simultaneous use of multiple project delivery methods. *Journal of Management in Engineering, 16*(3), 58–67.

Oke, A. E. (2016). Effect of bond administration on construction project delivery. *Organization, Technology and Management in Construction Journal, 8*(1), 1390–1396.

Oke, A. E., Olatunji, S. O., Awodele, A. O., Akinola, J. A., & Kuma-Agbenyo, M. (2016). Importance of team roles composition to success of construction projects. *International Journal of Construction Project Management, 8*(2), 141–152.

Olawale, Y., & Sun, M. (2010). Cost and time control of construction projects: Inhibiting factors and mitigating measures in practice. *Construction Management and Economics, 28*(5), 509–526.

Ren, Z., Kwaw, P., & Yang, F. (2012). Ghana's public procurement reform and the continuous use of the traditional procurement system: The way forward. *Built Environment Project and Asset Management, 2*(1), 56–69.

Shahu, R., Pundir, A. K., & Ganapathy, L. (2012). An empirical study on flexibility: A critical uccess factor of construction projects. *Global Journal of Flexible Systems Management, 13*(3), 123–128.

Shenhar, A. J., Dvir, D., Levy, O., & Maltz, A. C. (2001). Project success: A multidimensional strategic concept. *Long Range Planning, 34*(6), 699–725.

Tabish, S. Z., & Jha, K. N. (2011). Identification and evaluation of success factors for public construction projects. *Construction Management and Economics, 29*(2011), 809–823.

Takim, R., & Akintoye, A. (2002). Performance indicators for successful construction project performance. In D. Greenwood (Ed.), *18th Annual ARCOM Conference. 2*, pp. 545–555. Northumbria: Association of Researchers in Construction Management.

Tatum, C. B. (2011). Core elements of construction engineering knowledge for project and career success. *Journal of Construction Engineering and Management, 137*(10), 746–750.

Tenah, K. A. (2011). Project delivery systems for construction: An overview. *Cost Engineer, 43* (1), 30–36.

The American Institute of Architects. (2007). *Integrated project delivery: A guide.* California. Retrieved December 5, 2016, from http://www.info.aia.org/.

The Associated General Contractors of America. (2003). *Guidelines for a successful construction project.* Washington: The Associated General Contractors of America, the American Subcontractors Association, and the Associated Specialty Contractors. Retrieved September 14, 2016, from https://www.discountpdh.com/course/guideline-on-general-contractor-subcontractorrelations.

Thomson, D. (2011). A pilot study of client complexity, emergent requirements and stakeholder perceptions of project success. *Construction Management and Economics, 29*(2011), 69–82.

Toor, S. U., & Ogunlala, S. O. (2010). Beyond the 'iron triangle': Stakeholder perception of key performance indicators (KPIs) for large-scale public sector development projects. *International Journal of Project Management, 28*(3), 228–236.

Toor, S. U., & Ogunlana, S. O. (2008). Critical COMs of success in large-scale construction projects: Evidence from Thailand construction industry. *International Journal of Project Management, 26*(4), 420–430.

Touran, A., Gransbverg, D. D., Molenaar, K. R., Ghavamifar, K., Mason, D. J. & Fithian, L. A. (2009). *A gudebook for the evaluation of project delivery methods.* Washington: Transportation Research Board. Retrieved November 14, 2016, from http://www.national-academies.org/trb/bookstore.

Yu, A. G., Flett, P. D., & Bowers, J. A. (2005). Developing a value-centred proposal for assessing project success. *International Journal of Project Management, 23*(6), 428–436.

Zanjirchi, S. M., & Moradi, M. (2012). Construction project success analysis from stakeholders' theory perspective. *African Journal of Business Management, 6*(15), 5218–5225.

Chapter 6
Sustainability in Construction

Abstract The world is moving towards sustainable development and the construction sector is not being left behind because it plays a vital role in the development and economy growth of any country. However, construction processes have evolved over the years owing to several factors such as constant changes in the demands of stakeholders who are mostly clients, owners and regulatory agencies of infrastructural developments, and the core of construction is to provide value for money for clients and ensure they are duly satisfied. This has necessitated the alteration of the traditional ways of construction and ways of measuring the success of construction projects. This paradigm shift in construction and its related activities has given rise to sustainable construction which is geared at providing projects that are environmentally, socially, and economically beneficial to stakeholders, including the current and future generation within the immediate and distant environment of the project. This chapter commences by discussing the history of sustainable construction through the lens of sustainable development before providing a detailed explanation of the concept of sustainable construction. Various attributes and elements of sustainable construction are highlighted and discussed, followed by explanations of the levels of adoption of sustainability in the construction industry. Based on the identified elements, various barriers, drivers and benefits of sustainable construction are highlighted and explained.

Keywords Construction project · Project stakeholders · Sustainable construction · Sustainable development · Sustainable goals

Introduction

As discussed in Chap. 5, the construction industry plays a vital role in the development and economic growth of a country. However, construction processes have evolved over the years owing to several factors such as constant changes in demands of stakeholders who are mostly clients, owners and regulatory agencies of infrastructural developments, and the core of construction is to provide value for

A.E. Oke and C.O. Aigbavboa, *Sustainable Value Management for Construction Projects*, DOI 10.1007/978-3-319-54151-8_6

money for clients and ensure they are duly satisfied. This has necessitated the alteration of the traditional ways of construction and ways of measuring the success of construction projects. This paradigm shift in construction and its related activities has given rise to sustainable construction which is geared at providing projects that are environmentally, socially, and economically beneficial to stakeholders, including the current and future generation within the immediate and distant environment of the project.

This chapter commences by discussing the history of sustainable construction through the lens of sustainable development before providing a detailed explanation of the concept of sustainable construction. Various attributes and elements of sustainable construction are highlighted and discussed, followed by explanations of the levels of adoption of sustainability in the construction industry. Based on the identified elements, various barriers, drivers and benefits of sustainable construction are highlighted and explained.

History of Sustainability in Construction

One of the major global issues confronting the generality of people is climate change as a result of the effect of global warming. This is largely as a result of ozone layer depletion due to carbon emissions of harmful gases from the activities of man, such as manufacturing, and construction. Other contributing factors are rapid industrialisation, globalisation and collaboration, innovative practices to further satisfy customers and users or products, advancement in technology in dealing with aspect of human endeavour, excessive urbanisation and emigration to developed countries due to war and political instability, an increase in population, and an increase in resource utilization to meet the demands of populations among others.

These negative effects led to the call for sustainable activities that harmonise and maintain the balance of achieving developments that are economically profitable and favourable, socially beneficial to people and environmentally friendly to society at large. This is known as sustainable development which is geared at meeting the needs of the present without jeopardizing or compromising the ability of future generations to meet their own needs (Brundtland et al. 1987; Bourdeau 1999; Alwan et al. 2017). The construction industry is not in any way exempted from these global activities, practices and issues. However, the industry also has some peculiar attributes including the increasing complexity of projects, the involvement of stakeholders with diverse vested interests, more use of sophisticated equipment, long production durations, and innovation in design and construction process. The adoption of principles of sustainable development in construction gave rise to sustainable construction, indicating that the latter is an aspect of the former.

The five key principles of sustainable development (Queensland Department of Public Works 2008) that are geared at improving the quality of life of people and the environment without comprising economical value both now and in the future are the following:

- Integration;
- Community involvement;
- Precautionary behaviour;
- Equity within and between generations; and
- Continual improvement.

The integration principle is concerned with the ability to interpolate various elements of sustainable development, which are economic, social, and environmental, in key decision making as it relates to the project under consideration. This implies that each of the three aspects is considered in proposing, appraising, and agreeing on project decisions.

Every development is expected to have impact on the lives of people and their environment in one way or the other (either an immediate, future or a combination). In view of this, there is a need for community participation through a needs and impact assessment before, during, and after the construction process in order to involve the people and gain their support. This is more essential for government or public projects that involve a different group of stakeholders with diverse interests and that are designed for the good of the people. Moreover, most public projects are procured through tax payers' monies and for the public good of the citizenry.

The precautionary behaviour attributes of sustainable development are concerned with taking the necessary steps to limit and, if possible, prevent controllable threats to the people, the economy and the environment as a result of the development under consideration. Hence, necessary measures should be adopted at various stages to avert damage to the environment while considering the other two elements of sustainable development.

Sustainable development is not only concerned about the current quality of life of people who are affected by the development but also that of future generations. In view of this, one of the basic principles of sustainable development is to ensure equity within and between generations; this is termed the regenerative properties of the sustainable development projects. The basis is that development of today should not concern itself only with the current situation but should look ahead to its contribution to people later on.

The last principle which is related to continual improvement is the principle of taking action that will help to improve the environment and quality of life of the people without compromising the economic aspect, not only for the present but always. This is as a result of the declining nature of the environment which gave rise to sustainable development in the first instance.

There has been an increasing need for projects that are not only delivered according to the traditional measures of project success, that is, cost, time, and quality but construction projects that incorporate these measures (and other variants and emerging ones as discussed in Chap. 5) with little or low impact on the current and future environment as well as on the quality of life of the people. The need to incorporate standards and measures of project success relating to these issues necessitated the adoption of sustainable construction. This concept embraces an array of terminologies such as green construction, green housing, sustainable housing, and sustainable infrastructure. However, the term 'sustainable construction' is believed to incorporate all aspects of green buildings and construction as well as other sustainability issues.

Construction Sustainability

The quest for improved infrastructure and more sustainable built environment is on the increase (Brundtland et al. 1987; Bourdeau 1999; Tatum 2011; Marhani et al. 2012; Gan et al. 2015; Chang et al. 2016; Kibwami and Tutesigensi 2016; Alwan et al. 2017). In view of this, there has consequently been a widening of the scope of construction projects throughout the preconstruction, construction, and post-construction phases.

Sustainable construction is geared towards achieving the required measures of project success and performance with the least unfavourable environment impact, while encouraging cultural, social, and economic improvement at local, national, and global level (Häkkinen 2007). Sustainable construction entails the practice of adopting efficient, effective, improved, and healthier principles in the planning, construction, use, operation, maintenance, re-use, and eventual demolition of infrastructures with the emphasis on optimizing construction resources for a better environment and improved quality of life of the people for whom the specific development is meant. There are guiding principles to adopting sustainable practice which are also applicable to the construction industry (United Nations Environment Programme 2006; Kibert 2008; AlSanad 2015; Chang et al. 2016). These guiding principles include the following:

- Putting present and future prople at the centre of construction development;
- Applying a whole-life principle that entails a long-term perspective;
- Knowing and adopting cost-benefits analysis for investment appraisal of project;
- Having an open and supportive economic system in relation to political factors;
- Applying the principles of poverty eradication through job opportunities;
- Applying the principles of social inclusion of various concerned and affected parties;
- Understanding environmental limits and respecting them in the design and construction of projects;
- Adopting precautionary principles in construction processes;
- Adopting scientific knowledge in evaluating and achieving sustainable goals for construction projects;
- Applying the principles of transparency, knowledge management, and access to justice before, during, and after construction; and
- Applying the principle of making the polluter to pay using approaches such as carbon trading and resources usage monitoring.

Sustainability is being practiced in every developmental sector of the economy, including the construction industry. To differentiate sustainable construction from other sectors, various attributes have been highlighted by various authors, agencies, experts and other concerned stakeholders (United Nations Environment Programme 2006; Bal et al. 2013). These characteristics of sustainable construction include the following:

- Considering building regulations, legislations and standards;
- Designing and continuously maintaining the project to optimize its life span;
- Considering environmentally related issues inclusive of long and short aspects;
- Gearing government policies and incentives towards the awareness, adoption, and practice of sustainable construction practice; and
- Stakeholders (sponsors, financiers, end-users, buyers, and property developers) playing an active role in the encouragement of the practice of sustainable construction.

Attributes of Sustainable Construction

Sustainable construction should meet certain objectives such as customer satisfaction, user satisfaction, improved quality of life, flexibility for user changes, minimal use of resources, and support for desirable natural and social environments. For a construction project to be termed sustainable, the basic principles of sustainable construction should be used as a guide (Kibert 2008; Rafindadi et al. 2014). The seven principle of sustainable construction project should include the following:

- Reduce: This is concerned with the reduction of the consumption of various construction resources, namely, man, materials, machines, money, and method/process.
- Re-use: This is the ability to recoup, reprocess and salvage construction materials.
- Recycle: This entails the use of materials that can be use again for other purpose.
- Nature: This is concerned with the consideration of the environment to protect nature.
- Toxins: This is the elimination of the use of toxic materials that are harmful to the environment and people.
- Economics: This is the application of whole-life costing principles and techniques in the choice and adoption of construction materials.
- Quality: The focus of sustainable construction should be quality of life and the environment with the emphasis on both now and the future.

In addition to these criteria, there are other checklists developed by the international and national agencies to evaluate the design, construction, and post-construction phases of construction projects in line with adherence to sustainable goals. For instance, the

Building Research Establishment Environmental Assessment Method (BREEAM) which has been acknowledged as the first environmental assessment tool, is in the form of a fact file known as the BREEAM fact file and contains four aims and six objectives (Building Research Establishment Environmental Assessment Method 2015; Lee and Burnett 2008; Whang and Kim 2015). In this rating system, credits are awarded to performance based on the stated aims and objectives, which are summed together to arrive at a total score with which to grade the projects. The grades range from outstanding to excellent, very good, good and pass.

The following are the guidelines as indicated by Leadership in Energy and Environmental Design (LEED) for a design process of a sustainable construction (Lee and Burnett 2008; Sabol 2008; United States Green Building Council 2008):

- Efficiency analysis to evaluate lighting systems, HVAC systems and building envelope;
- Daylighting Analysis;
- Evaluation of existing building re-use;
- Site selection and how to use the site sustainably;
- Design charrettes that can be adopted to set project goals, including environmental goals;
- Energy modelling that can be adopted to provide the most energy efficient building that meets the project's budget;
- Sustainable material selection in relation to recycled content, recyclability, use of local materials and adoption of low volatile organic compounds (VOC); and
- Commissioning of all systems to ensure proper installation and operation of the building systems.

The sustainable building tool (SBTool), which is a Green Building Challenge (GBC) assessment technique, has been acclaimed to be more comprehensive for assessing sustainable construction because it incorporates all aspects of sustainability including that of green building and construction (International Initiative for Sustainable Built Environment 2007; Whang and Kim 2015). The tool contains 138 individual criteria that are categorised into 25 aspects and seven issues. Unlike BREEAM, four scoring criteria are adopted which include +5 for best practice, +3 for good practice, 0 for minimum acceptable performance and −1 for deficient. Some of the SBTool criteria for sustainable construction include the following:

Social

- Visual privacy from the exterior in principal areas of dwelling units;
- Access to private open space from dwelling units;
- Integration of project with local community;
- Minimization of construction accidents;
- Access to direct sunlight from living areas of dwelling units; and
- Access for physically handicapped persons.

Economic

- Minimization of construction cost;
- Minimization of operation and maintenance cost;
- Commercial viability;
- Affordability of residential rental or cost levels; and
- Support of local economy.

Culture and heritage

- Compatibility of urban design with local cultural values;
- Relationship of design with existing streetscapes; and
- Maintenance of heritage value of existing facility.

Apart from the highlighted criteria, there are also criteria developed by some researchers and experts for various aspects and work sections of construction projects. For instance, Shen et al. (2007) discuss a construction project sustainability performance checklist for different stages of a typical construction project. The identified stages are preconstruction (inception, design), construction, and post-construction (operation and demolition). Moreover, there are criteria designed to evaluate and manage specific aspects of construction based on the identified elements of sustainable practice. For instance, Waris et al. (2014) developed sustainability criteria for the selection of onsite construction equipment while Li and Wang (2016) discuss necessary checklists for large earthwork projects.

Elements of Sustainable Construction

The elements of sustainability are also referred to as sustainability goals by some scholars, researchers and experts. The early elements of sustainability are related only to the environment but over the years, a tripod approach has been adopted by most authors and world agencies (Bal et al. 2013; Hussin et al. 2013; Marhani et al. 2013; Rafindadi et al. 2014; Mazhar and Arain 2015; Ruparathna and Hewage 2015; Chang et al. 2016; Abd Jamil and Fathi 2016; Kibwami and Tutesigensi 2016; Li and Wang 2016; Alwan et al. 2017). The essence is to achieve a balanced sustainable implementation in the construction industry (Whang and Kim 2015). The three elements in the tripod approach include the physical or environment, the economic or financial environment, and the social or cultural (socio-cultural) environment. However, in the examination of evaluation of sustainability criteria, authors such as Waris et al. (2014) and AlSanad (2015) have included a fourth dimension encompassing such aspects as technical or technological, and political or legislative. However, the principal elements of social, environment, and economic remain the major elements of sustainable development and other variables can be factored into any of the three.

Socio-Cultural

This is also known as the ethical, social or cultural aspect of sustainable development. This aspect is geared towards producing construction projects that enhance the quality of human life, satisfaction of needs and expectations of people as well as enhancing corporate social responsibility. It is the aspect of sustainable construction that is geared towards responding to the needs of the people and society at large at every stage of project development. It entails major social issues such as the following:

- Employment;
- Health and safety;
- Well-being and human comfort;
- Education and training;
- Partnership;
- Culture and heritage;
- Security;
- Accessibility;
- Universal design;
- Community service; and
- Service quality.

Environmental

This is related to the immediate and distant environment of the project under consideration, which underscores the use of the term 'physical' by some scholars. It involves construction practices and processes that minimize the usage of natural resources and reduce harm to the environment. The environmental element of sustainability is concerned with the following:

- Location of project;
- Material and resources selection (water and energy);
- Material and resources conservation;
- Waste management;
- Atmosphere;
- Land utilization;
- Pollution control;
- Indoor environmental quality;
- Ecological environment;
- Environmental regulations;
- Transport; and
- Management.

Economic

This is concerned with the economic and financial aspect of sustainable construction and considers the cost of executing sustainable construction from inception through completion, use, re-use or conversion, and demolition. It is the ability to increase profitability through making use of construction resources efficiently. The economic aspect of sustainable construction should address initial costs, maintenance costs, running costs, future modification costs, and community costs. The construction process, practice, and methods should be geared to ensure that projects make a profit and return on investment without compromising the needs of the people. Such a project should also minimize the lifetime maintenance cost of the project. In view of this, the major considerations for economic sustainability include the following:

- Competitiveness;
- Productivity and profitability;
- Value for money;
- Partnering;
- Project success;
- Knowledge management;
- Durability;
- Quality management;
- Whole-life costing;
- Construction cost;
- Operating and maintenance cost;
- Affordability of cost level;
- Image and reputation;
- Innovation;
- Research and development; and
- Support of local economy.

Adoption of Sustainable Construction

The social, economic, and environmental impact of the construction industry cannot be overemphasized. Globally, the industry contributes about 35% of global CO_2 emission, 30% of greenhouse gas production during construction, and 18% of emission of gases during transporting and processing construction materials. In addition, it consumes 40% of all natural resources extracted in industrialized countries, uses 40% of total energy produced, consumes 25% of all timber production, accounts for 16% of total water consumption, consumes 40% of all raw materials, and generates 45–65% of waste deposited in landfills (Venkatarama

Reddy and Jagadish 2003; Wang et al. 2005; Keysar and Pearce 2007; Yudelson 2008; Sun 2011; AlSanad 2015).

For instance, in the Unites States of America, building alone accounts for the consumption of 40% of raw materials, 38% of CO_2 emissions, 46% of sulphur dioxide emission, 39–40% of energy use, 72% of electricity consumption, 12–14% of potable water use, and 30–35% of non-industrial waste output (Sabol 2008; Kam Shadan 2012). These statistics underscore the negative impact of construction activities on the present life of people and the environment. However, the future impact of these activities on the life of the people and the environment in general will be serious and damning.

The concept of sustainable construction has been widely adopted in most developed and a very few developing countries, while it is still gaining popularity and awareness in most developing ones. In the countries where it has become a benchmark for the delivery of construction projects, several sustainable assessment systems have been introduced to regulate and monitor sustainable construction activities (Sabol 2008; Hussin et al. 2013; Whang and Kim 2015). These assessment methods have been adopted by other countries with little or no amendments to regulate and control construction activities for the purpose of achieving sustainability. These include the following, among others:

- United States of America—Leadership in Energy and Environmental Design (LEED)
- United Kingdom—BRE Environmental Assessment Method (BREEAM)
- South Africa—Sustainable Building Assessment Tool (SBAT)
- Hong Kong—Building Environmental Assessment Model (BEAM) and Comprehensive Environmental Performance Assessment Scheme (CEPAS)
- Japan—Comprehensive Assessment Scheme for Built Environment Efficiency (CASBEE)
- European Union—Eco-Management and Auditing Scheme (EMAS)
- China—Evaluation Standard for Green Building (ESGB)
- Malaysia—Green Building Index (GBI)
- Sustainable building tool (SBTool)
- Green Building Certification Criteria (GBCC)

Barriers to Sustainable Construction

Despite the popularity and increasing awareness of sustainability goals, it is noteworthy that some barriers to its adoption still persist. However, the low level of awareness and adoption are more common in developing countries than in developed countries (Du Plessis 2007; Hussin et al. 2013; Djokoto et al. 2014; AlSanad 2015; Ametepey et al. 2015; Saleh and Alalouch 2015; Chang et al. 2016; Kibwami and Tutesigensi 2016). The barriers can be grouped together or classified into the various areas of sustainable development goals.

In discussing the barriers to sustainable construction, the initial tripod of physical, economic, and socio-cultural has been deemed insufficient, hence the addition of new concepts of professional, technical (technological), and political (legislative) factors especially for the developing economies of the world. For this aspect, four areas, namely socio-cultural, technological, economic, and political as identified by Du Plessis (2007) and Mousa (2015) have been adopted. The barriers are explained under each of the following groups (John et al. 2001; Gomes and Silva 2005; Majdalani et al. 2006; Du Plessis 2007; Djokoto et al. 2014; Ametepey et al. 2015; Gan et al. 2015; Mousa 2015; Saleh and Alalouch 2015; Chang et al. 2016).

Socio-Cultural Related Issues

This is concerned with the awareness and willingness of stakeholders to adopt sustainable construction principles contrary to their accustomed traditional method. The major barriers to socially sustainable construction include the following:

- Wrong perception of natural resources;
- Wrong perception of sustainability;
- Wrong perception of value management;
- Wrong perception of quality of life;
- Wrong perception of value research;
- Wrong perception of cost with focus on initial cost;
- Non-sustainability of current construction practices;
- Lack of consumer and public awareness;
- Lack of government incentives;
- Absence of media or communication channel;
- Illiteracy;
- Population competing for same resources;
- Lack of professional training;
- Attitude of common work culture;
- Not embracing corporate social responsibility; and
- Resistance to change from traditional construction processes.

Technological Related Issues

Another area of barriers is related to the technologies in term of manpower, materials, and equipment that are necessary for the achievement of sustainable construction. The lack of knowledge of sustainable technologies by professionals and other stakeholders in respect to the identified items constitutes major barriers to sustainable construction. The technological barriers are related to the following:

- Lack of alternative materials and technology;
- Importing new materials and technology;
- Lack of training on new materials and technology;
- Uncertainties in sustainable technological performance;
- Not embracing new materials and technology in building codes;
- Lack of local research and development and innovation;
- Non-utilization of natural resources;
- Misunderstanding of sustainable construction technological specifications; and
- Lack of adequate sustainable technological operation.

Economy-Related Issues

This is concerned with economic growth and the major sustainable construction barriers for this aspect are related to the following:

- Available natural resources that are not sustainable;
- Lack of alternative materials;
- Local competitiveness;
- Universal competitiveness;
- Lack of market control;
- Importing sustainable materials;
- Importing new material and technology;
- Non-availability of business information;
- Local market practices; and
- Monopoly of the construction business leading to a lack of innovation.

Politically Related Issues

This is inclusive of legislative barriers as well political stagnancy and instability. The identified barriers are concerned with the following:

- Lack of transparency;
- Lack of legislative changes and restructuring;
- Business environment that is not conducive;
- Lack of sustainable legislations and building codes;
- Weak law enforcement;
- Protection of investment;
- Non-sustainability of current construction permitting process;
- Political instability;
- Lack of political vision and will; and
- Bureaucracy in governance and administration.

General Barriers

In most economies, especially the developing ones, the predominant use of concrete of various forms—whether mass or reinforced, on-site or off-site, ready-mixed concrete (RMC) or site-mixed-concrete (SMC) – for construction has raised a major concern as a result of various disadvantages associated with it. In practice, the tilt-up method, which is concerned with the adoption of ready-mixed or off-site concrete, has proven to be a viable, effective, efficient, and globally acceptable sustainable concrete system (Saleh and Alalouch 2015; Aigbavboa et al. 2016; Choi et al. 2016). The method allows for a low waste of material, faster delivery of project, structural integrity, durability, and limited maintenance of finished products, and the enhancement of cost reduction as well as the achievement of value for money and satisfaction for project clients.

This method of construction (tilt-up method) is still gaining popularity in most developing countries and the relevance has not been fully utilized. This has resulted in an increase of associated problems. According to Mousa (2015), the major sustainable issues relating to the use of concrete include the following:

- Predominant use of site-mixed-concrete against ready-mixed-concrete;
- Limited presence and use of higher quality concrete;
- Scarce use of cement supplementary materials and blended cements;
- Lack of legislative and regulatory support for the use of ready-mixed-concrete;
- Lack or enforcement of licensing for engineers and associated professionals;
- Lack of transparency in the approval and permitting process;
- Deliberate poor selection and mixture of concrete content;
- Non-compliance with construction codes, regulations and specifications;
- Poor supervision and other construction practices;
- Interference by non-professionals in technical decision making;
- Market monopoly of cement production, thereby controlling the quality; and
- Lack of an environmental impact assessment and feasibility study.

Based on the views from various countries and the perception of stakeholders on the barriers to the adoption of sustainable construction (AlSanad 2015; Ametepey et al. 2015; Choi et al. 2016), the following are the major factors influencing the practice:

- Wrong perception of sustainable construction practice;
- Lack of awareness and understanding of the concept of sustainable construction;
- Lack of favourable economic conditions;
- Risks and uncertainties of implementing the concept of sustainable construction;
- Belief that sustainable construction is expensive and costly;
- Lack of clear benefits of sustainable construction;
- Lack of professional capacity;
- Lack of interest by developers to undertake sustainable construction;
- Unwillingness of stakeholders to change from traditional practices;
- Lack of expertise and qualified staff to execute sustainable projects;

- Lack of regulations and rules to promote the adoption of sustainable construction; and
- Lack of government support.

Drivers of Sustainable Construction

It has been explained earlier that sustainable construction has not been fully practised and adopted in most construction industries, especially in the developing countries. There are several factors responsible for the challenges while the major ones are related to stakeholders' awareness and knowledge of sustainable construction as well as lack of willingness to adopt sustainable practices (AlSanad 2015; Gan et al. 2015; Ruparathna and Hewage 2015; Chang et al. 2016; Kibwami and Tutesigensi 2016). In view of this, the drivers were discussed from the perspective of key stakeholders in the construction industry which include clients, governments, contractors, consultants and suppliers.

Clients

These include an individual, corporate body, or government that commissions construction projects for a particular need. They are grouped into public and private clients. The expected actions from clients include the following:

- Awareness of sustainable construction;
- Willingness and acceptance to adopt sustainable construction;
- Understanding of the concept of sustainable construction;
- Realisation that cost and profit should not be the only measure of project success; and
- Acceptance of sustainable goals as a measure of project success.

Governments

The governments in this regard are inclusive of every arm of government, agency and board at local, national, continental, and international level that is tasked with the responsibility of managing, controlling, and regulating the construction of projects to ensure their sustainability. The following are expected from the government in the quest to drive sustainable construction:

- Preparing policy standards and guidelines for sustainable construction;
- Adhering strictly to building approval plans at the planning stage of construction;
- Adhering strictly to building standards and regulations during preconstruction, construction, and post-construction stages of projects;
- Regulating and monitoring construction activities to standard;
- Fostering the preparation or adoption of environmental assessment plan for construction projects under their domain; and
- Reviewing legislation and standards to meet current sustainability standard.

Contractors and Suppliers

This set of individuals includes main contractors and different forms of subcontractors who are involved or who use their expertise, experience, and resources in the actual construction works. These individuals are expected to contribute the following in driving sustainable construction:

- Preparing an environmental impact assessment for the proposed project;
- Supplying and using sustainable materials and other associated resources;
- Arranging for sustainable waste management practice;
- Adhering standards relating to health and safety, energy saving and other sustainable construction practices;
- Sensitising the public on potential threats and ways to combat them when and where required; and
- Training and retraining personnel on sustainable construction practices and means of adoption.

Consultants

These are professionals trained and experienced in a particular field or area of service who are engaged by clients or owners to ensure the smooth running of construction activities. The roles of these individuals include the following:

- Applying relevant standards in preparation of construction documents and discharge of services;
- Producing innovative designs and associated documents for sustainable construction;
- Painstaking planning to ensure project documents compliance with sustainable standard;

- Ensuring that sustainable conditions are well spelt out to the understanding of contractors;
- Checking and monitoring materials supplied to site for conformity to standard;
- Monitoring the construction process, equipment, labour, and materials compliance to approved standard; and
- Continually checking other requirements for sustainable construction before, during, and after construction as required.

In general, the following are the drivers for sustainable construction:

- Knowing and understanding the concept of sustainable construction by professionals;
- Willing to adopt sustainable construction practice by stakeholders;
- Formulating policies and guidelines by government and regulatory agencies;
- Recognizing the payback period of sustainable construction;
- Acknowledging the intangible benefits from sustainable construction;
- Making educational programmes available;
- Obtaining the commitment and involvement of clients;
- Having appropriate and adequate rules and regulations;
- Ensuring the availability of guidelines and standards;
- Providing economic incentives;
- Securing the cooperation of project stakeholders to achieve sustainable construction;
- Obtaining support from professional institutions; and
- Obtaining support from financial institutions and sponsors.

Benefits of Sustainable Construction

Sustainable construction is beneficial for the achievement of economically profitable, environmentally friendly, and socially beneficial construction projects. Based on the identified elements, the following are some drivers of sustainable construction (Saleh and Alalouch 2015; Chang et al. 2016; Kibwami and Tutesigensi 2016):

Socio-Cultural

The drivers of the social aspect of sustainable construction include the following:

- Alleviating poverty;
- Ensuring the operations of a development that is compatible with local needs;
- Educating and training to increase awareness;

- Acknowledging corporate social responsibility (CSR);
- Maintaining health and safety at workplace;
- Creating green jobs;
- Developing capacity and skills; and
- Increasing awareness of sustainable construction practice.

Environmental

The associated benefits for environmentally sustainable construction practices include the following:

- Reducing the use of energy during construction;
- Reducing the use of water during construction;
- Reducing the use of construction materials during construction;
- Optimising lifecycle energy use;
- Recycling products;
- Re-using products;
- Using renewable in preference to non-renewable materials;
- Minimising pollutants that cause environmental degradation;
- Supporting environmental labelling and voluntary rating schemes;
- Implementing environmental management practices during construction stage such as documenting requirements in contract specifications;
- Including environmental aspects in decision making during construction (e.g. buying greener materials);
- Developing comprehensive data bases; and
- Enforcing and complying with environmental regulations.

Economic

The associated sustainable construction benefits for economic-related activities include the following:

- Ensuring capital cost saving;
- Ensuring commercial viability;
- Increasing investment returns;
- Retaining skilled labour;
- Ensuring financial affordability for intended beneficiaries;
- Reducing running costs;
- Creating employment such as using labour intensive construction;
- Promoting competitiveness through advancing practices that advance issues of sustainability;

- Choosing environmentally responsible suppliers/contractors who demonstrate environmental performance;
- Offering incentives for those applying a sustainability measure (e.g. lower interest rates, tax exemption, etc.) and vice versa;
- Using local resources (e.g. materials and workforce) in construction; and
- Increasing productivity and performance.

Summary

The world is moving towards sustainable development and the construction sector is not being left behind. This chapter explained various aspects of sustainable construction as an aspect of sustainable development with a view to preparing a background for Chap. 8 which explains how value management can be adopted for sustainable construction.

References

Abd Jamil, A. H., & Fathi, M. S. (2016). The integration of lean construction and sustainable construction: A stakeholder perspective in analyzing sustainable lean construction strategies in Malaysia. *Procedia Computer Science, 100*(2016), 634–643.

Aigbavboa, C. O., Oke, A. E., & Thole, Y. L. (2016). Sustainability of tilt-up construction method. *Procedia Manufacturing, 7*(2016), 518–522.

AlSanad, S. (2015). Awareness, drivers, actions, and barriers of sustainable construction in Kuwait. *Procedia Engineering, 118*(2015), 969–983.

Alwan, Z., Jones, P., & Holgate, P. (2017). Strategic sustainable development in the UK construction industry, through the framework for strategic sustainable development, using Building Information Modelling. *Journal of Cleaner Production, 140*(2017), 349–358.

Ametepey, O. S., Aigbavboa, C. O., & Ansah, K. S. (2015). Barriers to successful implementation of sustainable construction in the Ghanaian construction industry. *Procedia Manufacturing, 3* (2015), 1682–1689.

Bal, M., Bryde, D., Fearon, D., & Ochieng, E. (2013). Stakeholder engagement: Achieving sustainability in the construction sector. *Sustainability, 6*(5), 695–710.

Bourdeau, L. (1999). Sustainable development and the future of construction: A comparison of visions from various countries. *Building Research & Information, 27*(6), 354–366.

Brundtland, G., Khalid, M., Agnelli, S., Al-Athel, S., Chidzero, B., Fadika, I., et al. (1987). *Our common future*. Oxford: Oxford University Press.

Building Research Establishment Environmental Assessment Method. (2015). *BREEAM information.* BRE Environmental Assessment Method. Retrieved June 12, 2015, from http://www.bre.co.uk/index.jsp.

Chang, R.-D., Soebarto, V., Zhao, Z.-Y., & Zillante, G. (2016). Facilitating the transition to sustainable construction: China's policies. *Journal of Cleaner Production, 131*(2016), 534–544.

Choi, S. W., Oh, B. K., Park, J. S., & Park, H. S. (2016). Sustainable design model to reduce environmental impact of building construction with composite structures. *Journal of Cleaner Production, 137*(2016), 823–832.

Djokoto, S. D., Dadzie, J., & Ohemeng-Ababio, E. (2014). Barriers to sustainable construction in the Ghanaian construction industry: Consultants' perspectives. *Journal of Sustainable Development, 7*(1), 134–143.

Du Plessis, C. (2007). A strategic framework for sustainable construction in developing countries. *Construction Management & Economics, 25*(2007), 67–76.

Gan, X., Zuo, J., Ye, K., Skitmore, M., & Xiong, B. (2015). Why sustainable construction? Why not? *An owner's perspective. Habitat International, 47*(2014), 61–68.

Gomes, V., & Silva, M. G. (2005). Exploring sustainable construction: Implications from Latin America. *Building Research Information, 33*(2005), 428–440.

Häkkinen, T. M. (2007). Sustainable building related new demands for product information and product model-based design. *Journal of Information Technology in Construction, 12*(2007), 19–37.

Hussin, J., Rahman, I. A., & Memon, A. H. (2013). The way forward in sustainable construction: Issues and challenges. *International Journal of Advances in Applied Sciences, 2*(1), 31–42.

International Initiative for Sustainable Built Environment. (2007). *About SBTool 07, SBC08.* International Initiative for Sustainable Built Environment. Retrieved June 4, 2015, from http://www.iisbe.org/iisbe/sbc2k8/sbc2k8-download_f.htm.

John, V. M., Agopyan, V. & Sjostrom, C. (2001). On agenda 21 for Latin American and Caribbean construbusiness: A perspective from Brazil. *Agenda 21 for Sustainable Construction in Developing Countries – First Discussion Document.*

Kam Shadan, P. E. (2012). *Construction project management handbook.* Washington: Federal Transit Administration. Retrieved September 14, 2016, from https://www.transit.dot.gov/sites/fta.dot.gov/files/FTA_Report_No._0015_0.pdf.

Keysar, E., & Pearce, A. R. (2007). Decision support tools for green building: Facilitating selection among new adopters on public sector projects. *Journal of Green Building, 2*(3), 153–171.

Kibert, C. J. (2008). *Sustainable construction: Green building design and delivery* (2nd ed.). Hoboken, NJ.: Wiley.

Kibwami, N., & Tutesigensi, A. (2016). Enhancing sustainable construction in the building sector in Uganda. *Habitat International, 57*(2016), 64–73.

Lee, W. I., & Burnett, J. (2008). Benchmarking energy use assessment of HK-BEAM BREEAMand LEED. *Building and Environment, 43*(11), 1882–1891.

Li, W., & Wang, X. (2016). Innovations on management of sustainable construction in a large earthwork project: An Australian case research. *Procedia Engineering, 145*(2016), 677–684.

Majdalani, Z., Ajam, M., & Mezher, T. (2006). Sustainability in the construction industry: A Lebanese case study. *Construction Innovation, 6*(2006), 33–46.

Marhani, M. A., Jaapar, A., & Bari, N. A. (2012). Lean construction: Towards enhancing sustainable construction in Malaysia. *Procedia—Social and Behavioral Sciences, 68*(2012), 87–98.

Marhani, M. A., Jaapar, A., Bari, N. A., & Zawawi, M. (2013). Sustainability through lean construction approach: A literature review. *Procedia—Social and Behavioral Sciences, 101* (2013), 90–99.

Mazhar, N., & Arain, F. (2015). Leveraging on work integrated learning to enhance sustainable design practices in the construction industry. *Procedia Engineering, 118*(2015), 434–441.

Mousa, A. (2015). A Business approach for transformation to sustainable construction: An implementation on a developing country. *Resources, Conservation and Recycling, 101*(2015), 9–19.

Queensland Department of Public Works. (2008). *Smart and sustainable homes design objectives.* Brisbane: Technology and Development Division, Department of Public Works. Retrieved June 4, 2015, from http://www.sustainable-homes.org.au.

Rafindadi, A. D., Mikic, M., Kovacic, I., & Cekic, Z. (2014). Global perception of sustainable construction project risks. *Procedia—Social and Behavioral Sciences, 119*(2014), 456–465.

Ruparathna, R., & Hewage, K. (2015). Sustainable procurement in the Canadian construction industry: Current practices, drivers and opportunities. *Journal of Cleaner Production, 109* (2015), 305–314.

Sabol, L. (2008). *Measuring sustainability for existing buildings.* Washington: Design + Construction Strategies, LLC. Retrieved November 14, 2016, from http://www.dcstrategies.net.

Saleh, M. S., & Alalouch, C. (2015). Towards sustainable construction in Oman: Challenges & opportunities. *Procedia Engineering, 118*(2015), 177–184.

Shen, L.-Y., Hao, J. L., Tam, V. W.-Y., & Yao, H. (2007). A checklist for assessing sustainability performance of construction projects. *Journal of Civil Engineering and Management, 13*(4), 273–281.

Sun, H. (2011). Implementing sustainable development in the construction industry: Constructors' perspectives in the US and Korea. *Sustainable Development, 19*(5), 620–628.

Tatum, C. B. (2011). Core elements of construction engineering knowledge for project and career success. *Journal of Construction Engineering and Management, 137*(10), 746–750.

United Nations Environment Programme. (2006). *Sustainable building and construction initiative.* United Nations Environment Programme. Retrieved September 14, 2016, from http://www.uneptie.org/pc/SBCI/SBCI_2006_InformationNote.pdf.

United States Green Building Council. (2008). *Leadership in energy and environmental design.* Washington: United States Green Building Council. Retrieved June 4, 2015, from http://www.usgbc.org/Displaypage.aspx?CategoryID=19.

Venkatarama Reddy, B. V., & Jagadish, K. S. (2003). Embodied energy of common and alternative building materials and technologies. *Energy and Buildings, 35*(2), 129–137.

Wang, W., Zmeureanu, R., & Rivard, H. (2005). Applying multi-objective genetic algorithms in green building design optimization. *Building and Environment, 40*(11), 1512–1525.

Waris, M., Liew, M. S., Khamidi, M. F., & Idrus, A. (2014). Criteria for the selection of sustainable onsite construction equipment. *International Journal of Sustainable Built Environment, 3*(2014), 96–110.

Whang, S.-W., & Kim, S. (2015). Balanced sustainable implementation in the construction industry: The perspective of Korean contractors. *Energy and Buildings, 96*(2015), 76–85.

Yudelson, J. (2008). *The green building revolution*. Washington: Island Press.

Part IV
Value Management and Construction Sustainability

Chapter 7
Value Management as a Construction Management Tool

Abstract Value management can be described as a management tool that can be adopted for overall project management of construction projects. Other construction management techniques such as risk management, lean construction and management, total quality management, knowledge management, total asset management, and safety management can be incorporated into a value management study instead of setting up various management teams to manage these aspects for the same project. This chapter explains the application of value management for other construction management methods, techniques, and practices in the construction industry. It highlights various areas of the construction management techniques and how their practice can be incorporated into value management study. However, it should be noted that for some specific reasons, the highlighted and discussed management techniques will be more effective and efficient for a better performance of construction projects when they are integrated with value management study for the same project.

Keywords Construction project · Knowledge management · Lean management · Project management · Risk management · Total quality management · Value management

Introduction

The concept of value management as a discipline has been explained in Chaps. 2 and 3. This chapter explains the application of value management for other construction management methods, techniques, and practices in the construction industry. The management techniques are introduced as a result of major attributes of construction projects that were discussed in Chapt. 4, for the purpose of realizing the specific and overall goals of construction projects. This was also discussed extensively in Chap. 5.

Each of the management techniques, including the old, new, and emerging ones, have their benefits and make positive contributions to a project success if

A.E. Oke and C.O. Aigbavboa, *Sustainable Value Management for Construction Projects*, DOI 10.1007/978-3-319-54151-8_7

adequately practised and adopted but there are also some disadvantages if they are not well planned and implemented. One of the problems facing construction projects, especially those that are more complex with many stakeholders with vested interests, is the application of the various and many construction management principles to a single project. The problems associated with this practice in the construction industry include the following:

- More complexity of the construction process as a result of trying to incorporate all recommendations from each of the practices;
- Non-suitability of some of the practices due to the nature of the project, type of client and requirement, and societal issues;
- Need for adequate management of each practice;
- Need for technicalities from participants, especially the designers and other consultants, for incorporating their recommendations;
- Low level of or lack of understanding of some of the practices by designers and other consultants;
- Lack of willingness by designers and other consultants to accept change to their original design and documents;
- Cost implications of each process to clients, owners, sponsors and financiers;
- Repetition and contradiction in recommendations from different practices;
- Dilemma of which recommendation to abide with when there is contradiction;
- Requiring project manager or lead consultant to be able to manage more complex processes without compromising original duties and responsibilities;
- Requirements such as legislations for the enforcement of the recommendations may not be available;
- Technology, technicalities and methods to implement some of the recommendations may not be available or too expensive;
- Plant and equipment to implement some of the recommendations may not be available or too expensive (and possibly not economical);
- Materials (raw or finished product) to implement some of the recommendations may not be available or too expensive;
- Experts or people to implement some of the recommendations may not be available or too expensive;
- Problems with the incorporation of the recommendations may lead to a loss of faith in the practice by clients and original project design team; and
- Problems with the incorporation of the recommendations may result in construction issues leading to disputes, and conflict among project stakeholders. These may result in legal issues and project abandonment, which may subsequently result in cost and time overruns which are the major reasons for the initial introduction of the practices.

From the highlighted issues, it can be deduced that it is not wise to implement so many construction management techniques for the same project. It is therefore necessary to adopt a few that are able to incorporate principles of other methods for unified and integrated recommendations that will be simple to understand, easy to adopt, and beneficial to the project.

The concept of value management which is geared at incorporating teamwork and a systematic approach to attain the best function of projects at the lowest possible cost, is not a standalone practice. In fact, many other management principles in the areas of planning, managing, and controlling construction resources, namely, man, materials, machine, and methods, are incorporated at different phases of a value management study. This chapter therefore discusses how some of the management practices can be incorporated into a value management study.

Value Management as Project Management Tool

Project management has become a common practice for all forms of projects, especially capital projects. The objective of the practice is to execute projects so that the associated deliverables can meet the scope requirements relating to cost and time, and at acceptable quality, safety, risk and security levels in order to achieve the satisfaction of stakeholders with the end product (Treasury Government Office 2002; Gluch 2009; Kam Shadan 2012). The process involves identifying the user requirements, highlighting the project constraints, emphasizing resource needs, and establishing realistic objectives to meet the strategic goals of the project.

For typical construction project management, there are ten knowledge areas that should be considered in order to manage projects properly (Project Management Institute 2008; Banaitiene and Banaitis 2012). These include the following:

- Integration management;
- Scope management;
- Time management;
- Cost management;
- Quality management;
- Human resource management;
- Communication management;
- Risk management;
- Procurement management; and
- Stakeholders management.

As indicated in Chap. 5, there are numerous objectives and measures of construction project success and they often compete with each other. Therefore there is the need for a project manager who is skilful and experienced enough to be able to balance the various objectives and activities of all stakeholders and participants of the project. A project manager may be an independent individual or an organisation whose duties, responsibilities, and line of service are strictly in the area of managing projects. In some cases, other consultants such as an architect, engineer, or quantity surveyor can be appointed to shoulder project management responsibilities strictly as a project manager on the project, or sometimes to combine it with their primary professional responsibilities. This is common in countries where project management is not recognised and acknowledged as a discipline and any of the construction professionals can take over the duties of a project manager as part of their services. Project management of a construction project entails the overall and complete planning, controlling, organising, monitoring, and coordinating of a project throughout the project lifecycle (from conception through inception, completion, use, re-use, and demolition, as discussed in Chap. 4) with the purpose of meeting the needs of the client or owner as well as other requirements of cost, time, quality, satisfaction, sustainability, and other measures of project success.

Every construction project undergoes a series of management of resources from inception through completion, re-use, and demolition. However, project management in this context involves the appointment of a 'project manager' (or project management organisation) as a member and leader or coordinator of the design team who reports directly to the client or owner of the project. Project management cuts across all the stages of projects against value management that is introduced at one stage, preferably at the preconstruction phase, though the recommendations from the latter are useful for the entire lifespan of the project. Value management as a practice or study cannot replace project management as a process, but it should be considered as an aspect by project management organisations in discharging their duties and responsibilities. This will help to maximize the benefits inherent in value management for the better delivery of construction projects. It will also aid the proper implementation of value management recommendations regarding the project.

Value Management as a Risk Management Tool

Risk management is one of the aspects of project management and is concerned with the planning, controlling, and managing of risks and their potential threats and opportunities. Risk is the combination of the probability of uncertain events and their consequences. It is an uncertain event, condition, phenomenon, or circumstance that has the potential of affecting one or more aspects of the project objective negatively or positively (Tah and Carr 2000; Wang et al. 2004; Adams 2008; Project Management Institute 2008; Washington State Department of Transportation 2010; Banaitiene and Banaitis 2012; Odeyinka et al. 2012). The positive effects are known as opportunities while the negative consequences are referred to as threats. This implies that risks have the potential of influencing all the measures of project success including cost, time, quality, satisfaction, and sustainability, which may lead to conflict and disputes, including the project abandonment in some cases.

Risk Management Processes and Value Management

Every management practice has laid-down principles and processes that guide its activities, although with some necessary modifications, depending on specific factors. Risk management in project management involves four main processes, although this categorisation may vary from one scholar to the next owing to a combination or division of some of the processes (Carr and Tah 2001; Project Management Institute 2008; Banaitiene and Banaitis 2012). The processes include the following:

- Risk identification;
- Risk assessment;
- Risk mitigation; and
- Risk monitoring.

Risk management planning has been identified by some authors as the first phase of risk management before the identification stage. It involves the process and technique of deciding on how to exercise risk management activities from the inception and throughout the lifecycle of construction project (Washington State Department of Transportation 2010). It is a systematic process intended to maximize potential opportunities and minimize or eliminate potential threats.

Risk Identification

At the identification stage, the required exercise is to identify the type and source of risks inherent in the construction project under consideration. At the point of identification, construction risks can be classified as preconstruction, construction, post-construction or all phases – external or internal; local or international; technical, external, organisational, environmental, or project management; generic, project, business, safety, technical, or security; economical, environmental, social, or technological; and resources, operating, market, financial, or force majeure (Tah and Carr 2000; Wang et al. 2004; Mead 2007; Adams 2008; Washington State Department of Transportation 2010; Banaitiene and Banaitis 2012; Kam Shadan 2012; Amusan et al. 2013). However, regardless of the classification, the common risks in construction physical, global, local, economic, technological or political. Other forms of risks can be classified under one or more of these.

Risk Assessment

After identification of the risks, the next phase is a risk analysis which involves both the quantitative and qualitative evaluation of the identified risks. The qualitative process involves the use of brainstorming, the Delphi method, interviews, and checklists using a subjective approach in measuring the risk impact and probability of occurrence. On the other hand, the quantitative method is more sophisticated and objective in nature, using such principles as probability theory, cost risk analysis, decision tree analysis, fault tree analysis, and Monte Carlo simulation. Regardless of the adopted method, the basic steps for risk assessment include the following:

- Description of the risks and their nature;
- Project activities that may be impacted by the risk;
- The probability of occurrence of the risks;
- The potential cost impact of the risk; and
- The potential time or schedule impact of the risk.

Risk Mitigation

This involves various risk responses and management planning techniques for treating risks in order to maximize opportunities that come with the risk and reduce the supposed threats. In construction, as in any other projects, risks can be avoided, transferred, accepted (or retained), or shared (or mitigated) and the adopted method will depend greatly on the type of risk, resource availability, experience of the stakeholders, the type and nature of the project, and the type of clients.

Risk Monitoring

This is the overall management and controlling of risks by comparing the planned and actual risks to the project. It involves the tracking of the identified risks at the identification stage for effective assessment and mitigation techniques. It is possible to modify the proposed response technique at one stage or another and some risks may even be discovered during the course of executing the project: it will be necessary to manage them as well.

Incorporating Risk Management into Value Management Study

Value management is a risk management tool: the study involves the incorporation of risk management techniques in the evaluation of alternatives. During the study, risk identification, assessment, and response measures are proposed while the post-study phase (usually during and after construction) allows for the monitoring and evaluation of the identified risks. Risk management can be incorporated into value management through the following means:

- At the evaluation stage when the generated ideas are evaluated based on prescribed basis and techniques;

- At the development phase when the best alternative ideas are developed for easy application with all the associated indices such as cost and time;
- Consideration of risk management as a separate phase of the value management study or included in one of the original phases;
- Incorporating risk management for discussion at every phase of the study; and
- Inclusion of risk identification and management as one of the objectives and the basis of selecting alternatives before subsequent recommendation to the project design team.

Steps for Managing Risks in Value Management Study

The steps involved in risk management (Treasury Government Office 2002; Banaitiene and Banaitis 2012; Kam Shadan 2012) were highlighted. However, the same processes can be followed in managing the risks of alternative ideas in value management:

- What are the risks involved in each of the alternatives?
- What are the risks associated with each of the identified functions?
- What type of risks are they? (short or long term)
- What are the risks of not providing the functions?
- What are the programme risks for providing each of the functions?
- What are the economic or financial risks for providing each of the functions?
- What are the schedule risks for providing each of the functions?
- What are the quality or standard risks for providing each of the functions?
- What are the social risks for providing each of the functions?
- What are the environmental risks for providing each of the functions?
- What are the technological risks for providing each of the functions?
- What are suitable methods for their assessment?
- What are suitable measures for their management?
- What are the monitoring techniques during and after construction?
- What are alternative plans that can serve as contingency measures in case of occurrence?

Value Management as Lean Management Tool

Lean construction (LC) is one of the construction management principles that have been gaining popularity and awareness by researchers, experts, professionals, and key stakeholders in the construction industry. However, the principles were used in the industry before the advent of the term 'lean construction' (Marhani et al. 2012). However, the challenge lies in the full understanding of the terminology.

Previous studies reveal that the principle of lean production (LP), which is a broad term for lean management in various industries, is to minimize waste while focusing on traditional measures of project success, namely, cost, time, and quality. For instance, Womack and Jones (1996) observed that the practice will help to minimize costs, especially the indirect ones, without compromising the quality standards while reducing the manufacturing cycle time. However, over the years, the principle of value has been included as one of the objectives of lean construction (Womack and Jones 2003; Blakey 2008; Jorgensen and Emmitt 2008; Marhani et al. 2013).

Principles of Lean Construction and Value Management

The principles and concepts of lean construction are based on those of lean production or manufacturing concepts (Koskela 1992; Murman et al. 2002; Marhani et al. 2012; Marhani et al. 2013; Abd Jamil and Fathi 2016). From the first application of lean production or management through to construction (Koskela 1992), eleven principles of lean construction were highlighted. These which include the following:

- Reducing cycle time;
- Reducing variability;
- Reducing share of non-value adding activities;
- Increasing output flexibility;
- Minimizing number of steps, parts and linkages;
- Increasing output value through systematic consideration of customer requirement;
- Building continuous improvement into the process; balance flow improvement with conversion improvement;
- Focusing control on the complete process;
- Increasing process transparency; and
- Benchmarking.

However, these principles have been harmonised for lean construction and the new set of principles and their relevance to value management are discussed.

Delight End User

This is also the focus of value management study as value is examined from the basis of function which relates to quality, standard, and the like for the satisfaction of the stakeholders and end-users. The objective is to satisfy end-users by focusing on the customers.

Benefit of Lowest Optimum Cost

Value management principles relate to the achieving of best function of the construction project at the lowest possible overall cost which is an advancement on this principle of lean construction.

Quality Focused

Quality is one of the elements of value which is the primary focus of value management study. In lean construction, wastes are minimized without compromising the quality of the final product.

Focus on Culture and People

Value management study does not only consider the satisfaction of stakeholders and their influence on the project but also adopts the principle of sustainability. The people who will be affected by the project are also considered in the evaluation and selection of alternatives.

Waste and Inefficiency Elimination

Waste in construction is one of the contributory factors to the issue of unnecessary cost which value management seeks to eliminate.

Continuous Improvement of Performance

The principle of value management considers whole-life costing of the project which includes initial, running, maintenance, and operational cost. The objective is to allow for continuous improvement of project performance until the product's eventual demolition.

One-Point Contact for Effective Coordination

Through the adoption of teamwork principles that incorporate key stakeholders of the project under consideration, a value management study is able to come up with recommendations that have the potential of being accepted by all stakeholders.

Concepts of Lean Construction and Value Management

With reference to the authors cited earlier, the basic concepts of lean construction as adopted from lean development are just-in-time; total quality management; business process re-engineering; concurrent engineering; last planner system; teamwork; value-based management; as well as occupation health and safety assessment series, among others.

Just-In-Time (JIT)

This is concerned with waste reduction and minimization through the continuous improvement of construction resources, namely, man, materials, machines, money, and the method. Value management is designed to adopt this principle of resource management, not just to eliminate or minimize waste, but also to optimize value.

Total Quality Management (TQM)

TQM is a management approach that is geared towards the achievement and improvement of the quality of products, systems, or project throughout the lifespan. In the construction industry, TQM is to ensure that all elements, parts, and components of a construction project, as well as the entire project, are delivered to acceptable standards based on statutory regulations and practices. The consideration of quality is one of the basic concepts of value in value management and during the study efforts are geared towards ensuring that quality and associated indices are not compromised.

Business Process Re-engineering (BPR)

This is the business aspect of lean construction that is meant to adopt integrated management thinking and actions for the fast delivery of quality goods and services at low cost. This is because customers and end-users are always interested in an affordable product and it is therefore necessary to design and redesign the organisational business plan at various times with the emphasis on their satisfaction. Value management adopts such variables as quality, customer satisfaction, user satisfaction, and client satisfaction as basic measures of function and all the efforts of the study are geared towards achieving these at the lowest possible cost.

Concurrent Engineering (CE)

Concurrent engineering is the incorporation and monitoring of design change through the design process of a project. For construction, it entails the management of alterations and modifications to the original idea of the project at the preconstruction phase so as to retain the objective of the project. The principle of value management is always introduced during the design stage and various changes are introduced into the original design to arrive at the optimum one. This is presented to the original project team for subsequent incorporation of recommendations emanating from the study of the construction project.

Last Planner System (LPS)

This concept of achieving lean construction is meant to increase productivity and decrease unpredictability mainly due to the social process by dealing with project variability and increasing the reliability of the commitment of team members. This is achieved by making planning a mutual exercise which is a key principle of value management.

Teamwork

Teamwork is one of the fundamental concepts of lean construction. In Chap. 3, value management was described as a team-based study that incorporate individuals (professionals and non-professionals) in various team roles for the purpose of working together to achieve a common objective. In the construction industry, the objective is to provide value for money and return on investment through the application of various speculative and evaluative principles.

Value-Based Management (VBM)

Value-based management classifies value into product and process value. The former is concerned with the value of the project to the customer or end-user which relates to their satisfaction of the project in meeting their needs and expectations. On the other hand, process value is the value of the project for workers and participants who provide services and goods and have a direct influence on the outcome of the success of the project. As the name suggests, this concept of lean construction is fundamental to value management study in that the whole-life situation of the project is considered which includes benefits to the participants and future benefits to the customers and end-users.

Occupation Health and Safety Assessment Series (OHSAS)

OHSAS is prepared to help the construction industry, especially the firms, to manage their occupational health and safety risks. This will help in improving the productivity of workers through an increase in commitment and job satisfaction. One of the measures of value and function in value management is health and safety and this concept of lean construction can be incorporated during the study.

Value Management as Total Quality Management Tool

Quality management should be ensured at the preconstruction, construction, and post-construction stages of any construction project. It involves quality assurance (QA) and quality control (QS) established before and during actual construction respectively. It is important at the preconstruction stage so that quality requirements are included and explained in various construction documents for tendering and the final contract. At the construction phase, it is necessary to highlight the procedure that should be followed in case of alterations or modifications to contract documents so as not to compromise the quality during the adjustment and amendments.

Procedures for Including Total Quality Management in Value Management

For total quality management to be efficient and effective in ensuring that construction projects are delivered to the right and acceptable standard, some necessary procedures of total quality management are required (Kam Shadan 2012). However, the same procedures are important in incorporating quality management principles into a value management study and the procedures should be observed for various alternative ideas. The procedures include the following:

- Checking of calculations, drawings, specifications and other documents by other members of value management team not directly associated with the preparation;
- Verifying of various drawing and design documents against the project scope;
- Checking of design and other documents for constructability;
- Checking compliance of design and other documents with operating and maintenance requirements;
- Checking compliance of design and other documents with statutory and regulatory requirements;
- Verifying of cost effectiveness of alternative designs;
- Checking of conformity of design and other documents to client need and user requirements; and
- Reviewing of design and other documents generally for other quality checks and compliance.

Benefits of Including Total Quality Management in Value Management

A value management study that incorporates total quality management is expected to be more productive and effective in helping clients and project owners to achieve their objective without compromising the value and function of various parts, elements, or the whole project. Significant improvement has been achieved by organisations, processes, and practices that embrace the principle of total quality management (The Associated General Contractors of America 2003). A value management study that embraces total quality management is expected to be beneficial in the following ways:

For construction projects and stakeholders

- Improvement in construction project performance;
- Improvement in the culture of the construction industry;
- Increased satisfaction of project owner, client or sponsor;
- Increased satisfaction of original project team;
- Increased satisfaction of customers and end users of construction project;
- Improved working environment for workers and other project stakeholders;
- Higher profit margin and savings for project clients and owners;
- Safer project execution and lower insurance cost;
- Fewer accidents and injuries relating to the construction work;
- Reduced rework and warranty work;
- Reduced variation and other forms of additional work; and
- Reduced effect of fluctuation on the construction project.

For value management team

- Greater efficiency in value management practice;
- Engagement and participation in meaningful value management study; and
- Improved reputation of the team.

For value management team members

- Greater self-satisfaction of value management team member;
- Higher morale of participants, that is value management team; and
- Provision of additional source of livelihood for members of the team.

Value Management as Knowledge Management Tool

Knowledge is fast becoming a key business concern and economic resource, especially in developed economies of the world (Edvinsson 2002; Ravishankar et al. 2011). This is as a result of the fact that well packaged knowledge is important capital to individuals, organisations, and processes. Moreover, innovation and advancement in technologies are as a result of knowledge gathering, sharing, and exploitation (The Associated General Contractors of America 2003; Oke et al. 2013). An important attribute of knowledge is that it can be transferred or shared with others for the purpose of teamwork effectiveness; therefore good, clear, effective and efficient communication among project stakeholders is a key ingredient for the success of construction projects.

Knowledge management involves the planning, controlling, and general management of information, data, and other forms of knowledge, whether tangible or intangible, in the quest of making it beneficial to a group of people or processes. The major process of knowledge management includes the following:

- Creating;
- Securing;
- Capturing or gathering;
- Organising or coordinating;
- Combining;
- Diffusing or distributing;
- Using; and
- Exploiting.

For effective and efficient knowledge management, these processes should be continuous and interwoven. This implies that there is no end to knowledge management; it is a cycle in that the creation of new knowledge commences as soon one is discovered and utilized. Moreover, new knowledge can be created at any of the stages, indicating that a stage is not independent of the others.

The construction industry as a whole and the process of producing projects involve the use of various data, information, evidence, and facts from diverse stakeholders and participants. The stakeholders are multidisciplinary; they are from formal (organised) and informal sectors and have a direct or indirect influence on construction projects. The formal sector stakeholders are usually organised and they include professionals such as project managers, architect, quantity surveyors (cost estimators), engineers, sponsors, financiers, statutory bodies, and other organised bodies or organisations tasked with a particular project responsibility. The informal sector stakeholders, who can be organised or unorganised, include such people as artisans, the community, and users. As a result of numerous stakeholders and the subsequent amount of information, knowledge management becomes an important issue in bringing together individual information or knowledge to form a corporate one that will be useful not just for the team, but for successful delivery of construction projects.

The importance of knowledge management to the construction industry cannot be overstressed. Despite its importance and the huge contribution to the economy of any country, the industry is still perceived as having low levels of performance and productivity. Also, there is a need to transfer and share knowledge of previous projects in the planning and actualisation of similar ones so as to avoid repeating the same mistakes and to build on existing knowledge for better performance. Another reason is the increasing demand for innovative projects that do not only meet the traditional measures of project success but are generally sustainable for current and future generations.

Knowledge management is fundamental to the success of any construction projects. The first stage of any project is the conception phase where the idea and need for a project arises in the mind of the client, who can be an individual, a group of people or an organisation. Right from the inception phase, it is expected that client will share information regarding the need for the project with an architect, project manager, or the lead consultant. The next stage is a feasibility and viability study which requires the gathering of further information to evaluate the possibility and profitability of the projects based on the initial idea of the client that must have been improved during the inception phase. This information is useful for every other stage, including design, planning, site work, handover, operation and usage, conversion, or re-use, and demolition. The level of knowledge exchange in the industry is multidimensional, interwoven, continuous, and cyclical in that information from one stakeholder may be used by several others and vice versa throughout the project lifecycle. For instance, in a typical building project, information in the form of drawings and sketches by an architect becomes a basis for engineering drawings and the two sets of information are useful for quantity surveyors in preparing cost documents such as the bill of quantities.

Despite the importance of knowledge and its management, it will not achieve its objective in construction project unless it is properly applied towards achieving a successful project with value for money and return on investment. Value management is a knowledge-based practice that gathers various forms of information geared towards achieving value in construction projects with the emphasis on function and cost in order of importance. Therefore, employing innovative problem-solving tools of knowledge management are inevitable if the efficiency of value management is to be enhanced. In fact, value principles, procedures, and stages as discussed in Chap. 3 indicate that value management is built around knowledge management.

Knowledge Management Requirements for Value Management

Several scholars have identified some factors necessary for a system to serve as knowledge management tool (Andawei 2001; Sharimllah et al. 2009; Dong-Gil and Alan 2011). The requirements are highlighted and explained in relation to the principle, practice, and process of value management.

Support of Knowledge Management Life Cycle

The life cycle consideration is one of the requirements of knowledge management. This has been discussed in the preceding section. Value management entails and adopts the principle of gathering, evaluating, and exploiting data for its activities.

Mechanism for Knowledge Validation

The presence of mechanisms and processes for the authentication and validation of knowledge is one of the requirements for effective knowledge management practice. However, this is the purpose of the evaluation and developmental phase of value management.

Ability to Integrate with Existing IT System

Another requirement of a knowledge management system is the ability to integrate with existing information communication technology systems within real or virtual settings. The practice of value management has been modernised through the use of a virtual team system against the traditional 40-h workshop and several ICT tools are now adopted for the evaluation of alternatives.

Flexibility and Ease of Use

Despite the order of activities and processes to be followed, one of the attributes of value management is flexibility in such aspects as the choice of team members, selection of venue, openness to ideas, and the use of alternatives.

Freshness of Knowledge

An aspect of value management is the collection of data for various aspects of construction projects, not only from previous projects or studies, but also through current market surveys and analyses. A forecast is also considered, especially in evaluating lifecycle cost of alternatives.

System Design

Design of system in knowledge management is expected to be in accordance to culture, business processes, and the goals of the project. In value management, essential knowledge is related to the needs and requirements of project client as well as other requirements relating to cost, time, quality, sustainability, satisfaction of various stakeholders, and health and safety compliance. These are adopted in the analysis and evaluation of alternatives in recommending the best alternative to the project team. Moreover, the team approach allows for the involvement of community participation in the study.

Barriers to Application of Knowledge Management in Value Management

Judging by the criteria highlighted and discussed above and the concepts and principles of value management, it could be concluded that knowledge management can be incorporated into value management study for construction projects. However, the following may hinder the application and effectiveness of knowledge management in construction value management; they should be considered and analysed in order to minimize or eliminate them:

- Unwillingness of individual to share knowledge;
- Lack of effective communication among stakeholders;
- Lack of funds for knowledge gathering and sharing;
- Difficulty in, and lack of policies for capturing knowledge;
- Unwillingness to change current operating systems;
- Lack of proper technical experience;
- Government policies;
- Lack of adequate and up-to-date data;
- Difficulty in locating knowledge;
- Difficulty in generalizing and storing data;
- General misunderstanding about knowledge management practice;
- Lack of a successful and applicable knowledge management model; and
- Lack of cooperation among stakeholders, especially professionals.

Summary

Value management can be described as a risk management tool that can be adopted for the overall project management of construction projects. Other construction management techniques such as risk management, lean construction and management, total quality management, knowledge management, total asset management, and safety management can be incorporated into value management study instead of setting up various management teams to manage these aspects for the same project.

This chapter highlighted various areas of the construction management techniques and how their practice can be incorporated into value management study. However, it should be noted that in some cases the highlighted and discussed management techniques will be more efficient and effective for a better performance of construction projects when they are integrated with value management study for the same project.

References

Abd Jamil, A. H., & Fathi, M. S. (2016). The integration of lean construction and sustainable construction: A stakeholder perspective in analyzing sustainable lean construction strategies in Malaysia. *Procedia Computer Science, 100*(2016), 634–643.

Adams, F. K. (2008). Construction contract risk management: A study of practices in the United Kingdom. *Cost Engineering, 50*(1), 22–33.

Amusan, L., Joshua, O., & Oloke, C. O. (2013). Performance of build-operate-transfer projects: Risks' cost implications from professionals and concessionaires perspective. *European International Journal of Science and Technology, 2*(3), 239–250.

Andawei, M. M. (2001). Application of network-based techniques in the cost control and management of construction works. *The Quantity Surveyors, 37*(4), 24–27.

Banaitiene, N. & Banaitis, A. (2012). Risk management in construction projects. In N. Banaitiene (Ed.), *Risk Management—Current Issues and Challenges* (pp. 429–448). London: Intech Open Science. Retrieved September 14, 2016, from http://cdn.intechopen.com/pdfs-wm/38973.pdf.

Blakey, R. (2008). *An introduction to lean construction.* Retrieved November 14, 2016, from http://www.touchbriefings.com.

Carr, V., & Tah, J. H. (2001). A fuzzy approach to construction project risk assessment and analysis: Construction project risk management system. *Advanced Engineering Software, 32* (10–11), 847–857.

Dong-Gil, K., & Alan, R. D. (2011). Profiting from knowledge management: The impact of time and experience. *Information Systems Research, 22*(1), 134–152.

Edvinsson, L. (2002). Knowledge is about people, not databases. *Industrial and Commercial Training, 31*(7), 262–266.

Gluch, P. (2009). Unfolding roles and identities of professionals in construction projects: Exploring the informality of practices. *Construction Management and Economics, October* (2009), 959–968.

Isle of Man. Treasury Government Office. (2002). *Procedure notes for management of construction project.* Douglas: Treasury Government Office. Retrieved September 14, 2016, from https://www.gov.im/media/383284/procedure_notes_for_management_of_construction_projects.

Jorgensen, B., & Emmitt, S. (2008). Lost in transition: The transfer of lean manufacturing to construction engineering. *Construction and Architectural Management, 15*(4), 383–398.

Kam Shadan, P. E. (2012). *Construction project management handbook.* Washington: Federal Transit Administration. Retrieved September 14, 2016, from https://www.transit.dot.gov/sites/fta.dot.gov/files/FTA_Report_No._0015_0.pdf.

Koskela, L. (1992). *Application of the new production philosophy to construction: Technical Report No. 72. CIFE.* Stanford University, California.

Marhani, M. A., Jaapar, A., & Bari, N. A. (2012). Lean construction: Towards enhancing sustainable construction in Malaysia. *Procedia—Social and Behavioral Sciences, 68*(2012), 87–98.

Marhani, M. A., Jaapar, A., Bari, N. A., & Zawawi, M. (2013). Sustainability through lean construction approach: A literature review. *Procedia—Social and Behavioral Sciences, 101* (2013), 90–99.

Mead, P. (2007). Current trends in risk allocation in construction projects and their implications for industry participants. *Construction Law Journal, 23*(1), 23–46.

Murman, E., Allen, T., Bozdogan, K., Cutcher, J. G., McManus, H., & Widnall, S. (2002). *Lean enterprise value: Insights from MIT's Lean Aerospace Initiative*. New York: Palgrave.

Odeyinka, H., Lowe, J., & Kaka, A. (2012). Regression modelling of risk impacts on construction cost flow forecast. *Journal of Financial Management of Property and Construction, 17*(3), 203–221.

Oke, A. E., Ogunsemi, D. R., & Adeeko, O. C. (2013). Assessment of knowledge management practices among construction professionals in Nigeria. *International Journal of Construction Engineering and Management, 2*(3), 85–92.

Project Management Institute. (2008). *Guide to the project management body of knowledge* (4th ed.). Newton Square: Project Management Institute.

Ravishankar, M. N., Pan, S. L., & Leidner, D. E. (2011). Examining the strategic alignment and implementation success of a KMS: A subculture-based multilevel analysis. *Information Systems Research, 22*(1), 39–59.

Sharimllah, D. R., Siong, C. C., & Hishamuddin, I. (2009). The practice of knowledge management processes: A comparative study of public and private higher education institutions in Malaysia. *Journal of Information and Knowledge Management Systems, 39*(3), 203–222.

Tah, J. H., & Carr, V. (2000). A proposal for construction project risk assessment using fuzzy logic. *Construction Management and Economics, 18*(4), 491–500.

The Associated General Contractors of America. (2003). *Guidelines for a successful construction project.* Washington: The Associated General Contractors of America, the American Subcontractors Association, and the Associated Specialty Contractors. Retrieved September 14, 2016, from https://www.discountpdh.com/course/guideline-on-general-contractor-subcontractorrelations.

Wang, S. Q., Dulaimi, M. F., & Aguria, M. Y. (2004). Risk management framework for construction projects in developing countries. *Construction Management and Economics, 22*(3), 237–252.

Washington State Department of Transportation. (2010). *Project risk management: Guidance for WSDOT projects.* Washington: Washington State Department of Transportation. Retrieved November 14, 2016, from http://www.wsdot.wa.gov/publications/manuals.

Womack, J. P., & Jones, D. T. (1996). *Lean thinking*. New York: Simon and Schuster.

Womack, J. P., & Jones, D. T. (2003). *Lean thinking: Banish waste and create wealth in your corporation*. New York: Simon and Schuster.

Chapter 8
Sustainable Value Management

Abstract Sustainability goals are becoming more important indices of construction project success. This is as a result of various planning and regulatory standards, especially regarding sustainability issues, that are currently in place for the control and monitoring of construction process and projects by national and international agencies. Moreover, owing to its popularity and necessity, clients and some other construction participants are also demanding its adoption as a major objective of construction projects. In view of this, this chapter explains value management as a construction sustainability tool in the development and delivery of sustainable construction projects. Various factors to be considered before initiating a sustainable value management study are discussed. The suitability of the study for a construction project is also explained by highlighting and discussing various forms of projects that will benefit significantly from its adoption. More so, a sustainable value management model incorporating cost, value, quality, and time around flexible sustainable goals is discussed and the model is related to three elements of sustainability, namely, economic, social, and environmental. Further to this, the necessities, principles, and stages of sustainable value management study are identified and explained accordingly.

Keywords Construction project · Sustainable construction · Sustainable development · Sustainable value management · Value · Value management

Introduction

Chapter 7 focused on how the principles of various construction management techniques can be incorporated into value management to enhance the contribution of the study for better performance of construction projects. Moreover, in Chaps. 5 and 6 it was mentioned that sustainability goals are becoming more important indices of construction project success. This is as a result of various planning and regulatory standards, especially regarding sustainability issues, that are currently in place for the control and monitoring of construction process and projects by

A.E. Oke and C.O. Aigbavboa, *Sustainable Value Management for Construction Projects*, DOI 10.1007/978-3-319-54151-8_8

national and international agencies. Moreover, owing to its popularity and necessity, clients and some other construction participants are also demanding its adoption as a major objective of construction projects. In view of this, this chapter explains value management as a construction sustainability tool in the development and delivery of sustainable construction projects.

Initiating Value Management Process

One of the current key issues concerning the construction industry is that construction projects are underperforming in terms of their capacity to deliver value to their customers (Cabinet Office 2011; Aghimien and Oke 2015; Jollands et al. 2015). To make construction projects more affordable to customers or clients, it is necessary for stakeholders in the industry to work towards eliminating inefficiency and waste, as well as stimulating a very high level of continuous innovativeness. Value management is a fundamental, effective, and efficient construction management tool that can be adopted to combat this problem in the construction industry. However, the following factors should be given attention before considering value management for construction projects:

- Type and cost of project;
- Size and complexity of project;
- Clients needs and project requirements (goals and objectives);
- Perceived potential for value improvement;
- Perception of client and/or other project team members to the study;
- Current stage or phase of project;
- Type of stakeholders to be involved who will be beneficial to the study;
- Availability of the stakeholders;
- Willingness of the stakeholders to participate; and
- Cost of value management study.

Suitability of Value Management for Sustainable Construction Projects

Value management has been mandated for specific types of projects based on cost, size, and type in various countries, including the USA, UK, Germany, France, Australia, Japan, Korea, India, and Malaysia. Realising the importance and benefits of the study, these countries have established various value organisations to plan, manage, and monitor value management study for various forms of project emanating from manufacturing and construction. For construction projects, value management is suitable for all forms of projects but the process and participants may have to be modified to suit the project under consideration. Value management is applicable, but not restricted to costly, complex, repetitive, new, budget-confined, and public projects.

Costly Projects

The principle of functional analysis as well identifying unnecessary costs in value management makes the practice suitable for expensive construction projects. Potentially, the adoption of value management can help to save up to 30% of construction cost without compromising function, quality, sustainable goals, or other requirements.

Complex Projects

A typical construction project is termed to be complex owing to the involvement of various stakeholders with various interests and the focus on achieving multiple goals at the same time. The more stakeholders and project goals, the more complex the project. Value management adopts a multidisciplinary teamwork approach and key stakeholders to the project can be well managed by including them in the value management team. This will allow for effective and efficient collaboration and discussion of project goals with the aim of arriving at a common ground.

Repetitive Projects

No two construction projects are the same, even if they are similar in shape and size, and are located within the same area or possess the same features. For instance, the ground conditions may not be the same; or users may be different which may affect other requirements. However, some basic features are common to repetitive works, including same design, scope, and objectives. Value management study can therefore be adopted to examine the basic features of the projects with the focus on maximizing their functions at the least possible cost. Hence, VM can be carried out for one of the projects and the recommendations adopted for the others.

New Projects

In this context, new projects are construction projects with a new type of design that requires the adoption of new technology, new materials, new equipment, or new knowledge from workers. In some countries, sustainable goals are new concepts and efforts at adopting them as part of construction projects objectives will term such projects as new because the goals have to be incorporated into the design. Owing to these factors, some unknown risks are deemed to be inherent in the project and value management can be adopted as a risk management tool for the purpose of gathering new information concerning various aspects of the project. This will help to understand the elements of the project, possible risks, and how to respond with appropriate measures before the commencement of the actual work on site.

Budget Restricted Projects

One of the key requirements at the inception stage of a project is the understanding of the source of funding for the project and the proposed budget for the project. It is possible the client has assigned a certain amount for the project and in this case, the onus is on the design team and other consultants to ensure that the project is planned in such a way that the overall cost of executing the project will not exceed the budget. Value management can be adopted for this purpose, whilst the project budget will be a key factor in the evaluation and adoption of the best alternatives and proposals for the project. The concept of unnecessary cost as a result of unnecessary materials and technology will also be a major focus of the study. At the end of the study, it is expected that proposals and recommendations by the value management team to the project team will be in line with the given budget while retaining functions.

Public-Inclined Projects

There are two categories of public-inclined projects. The first and most common are government or public projects that are designed and executed for the benefit and use of the public or people in general. The other category encompasses private, individual, or corporate projects that require the contribution of the public as a result of their effect on the quality of the public life and environment. For instance, the development and construction of a privately-owned cement manufacturing industry exploring limestone in a particular area will require that the people in the area—among other stakeholders—are included right from the planning and design of the industry. Value management can be adopted to bring the stakeholders to a common table for discussion and brainstorming with a view to uniting their opinions for the betterment of the people and the private investors. This is fundamental, not only to the successful planning and execution of the projects, but also for the effective and efficient usage and operation of the project after completion.

Sustainable Value Management Model

The identification and understanding of success measures are key factors for the success of any project, including construction projects. This becomes more important for public projects where multiple and diverse stakeholders with various interests are involved. It is becoming difficult to assess the success of construction projects owing to the lack of a universally accepted definition of project success and its measures (Abidin and Pasquire 2005; Tabish and Jha 2011; Shen and Yu, 2012; Noor et al. 2015). It is therefore necessary to recommend a method that will not only incorporate the basic measures of project success but that will help in ensuring that all construction projects comply with the minimum acceptable standards, regardless of their primary purpose.

Despite the emergence of new measures of project success, the importance of the traditional measures of time, cost, and quality cannot be neglected. In fact, most of the new measures stem from these primary variables but with some adjustment and modification to meet the current and future requirement of society at large. One of the new emerging measures of construction projects success is the achievement of sustainable goals as discussed in Chap. 5 and re-emphasized in almost every other chapter of this book. However, based on the explanation and discussion regarding value management, the adoption of the discipline for construction projects will not only aid the achievement of sustainable projects but will also help in the delivery of projects that optimize all the measures of project success. One of the fundamental benefits of value management is that it promotes adaptability and flexibility. The incorporation of sustainable construction issues into value management practice leads to what is termed 'sustainable value management' and this is described in Fig. 5.1.

There are existing models to achieve optimum success for construction projects, starting with the Iron Triangle by Atkinson (1999) as well as several others like those of Chan (2001), Takim and Akintoye (2002), Al-Yami and Price (2005) and Shahu et al. (2012) incorporating flexibility and effectiveness. The model in Fig. 5.1 incorporates virtually all the measures of project success identified in Chap. 5, with the adoption of 'flexibility' to allow participants to determine the goals as related to the particular project under consideration. The measures of cost, time, value, and quality are interrelated as discussed in Chap. 5 and the flexible goals are expected to be modified from one project to the other but nevertheless retain the basics. The basics in this regard are that all sustainable goals will be accustomed to a level of importance but this will vary depending on the nature and

Fig. 5.1 Sustainable value management for construction project

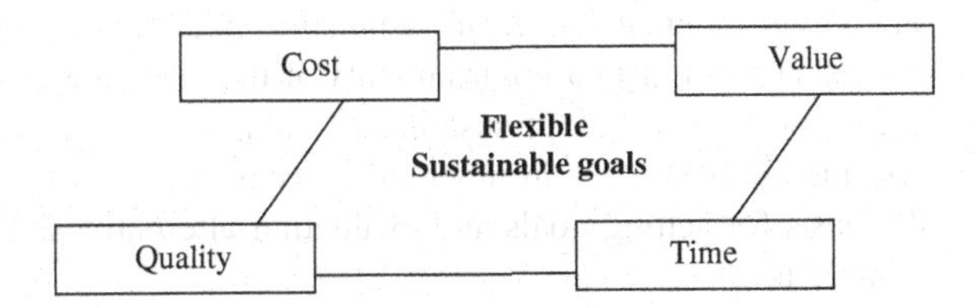

objectives of the project as well as the level of effectiveness required for each measure of project success.

Value Management for Financial Sustainability

This is the economic aspect of sustainable construction which is concerned with various forms of costs and financial implications associated with construction resources, materials, and the entire project. For sustainable value management, the following considerations of economic sustainability should be considered in the evaluation and selection of alternatives:

- Maximize the use of local and available materials;
- Maximize the use of local and available manpower;
- Maximize the use of local and available machines and equipment;
- Maximize the use of affordable materials;
- Maximize the use of resources with lower overall cost; and
- Minimize idle time by maximizing use of resources when needed.

Value Management for Social Sustainability

The social sustainability goal is concerned with the need to enhance the quality of life of people, meet the expectations of people, and satisfy their needs without compromising other aspects of the goal. For sustainable value management, the best alternatives should fulfil the following socially related functions:

- Ability to develop capacity and skills of the people;
- Ability to alleviate poverty among people;
- Ability to guarantee health and safety of people during construction and usage;
- Ability to meet local needs; and
- Ability to enhance corporate social responsibilities.

Value Management for Environmental Sustainability

The environmental aspect of sustainable goals is geared towards ensuring that the immediate and distant environments of the project are friendly now and in time to come. In order to adopt sustainable value management for this aspect, the major considerations for environmentally sustainable construction practice should form the basis for setting goals and evaluating alternatives. The best alternatives should be able to:

- minimize the use and consumption of resources, especially raw materials such as water and energy, during construction;
- minimize the use and consumption of resources during the project usage;
- minimize the use and consumption of resources during the conversion and re-use phase;
- minimize the use and consumption of resources after the expiration of lifespan, that is, at demolition stage;
- maximize the use of recyclable, renewable and re-usable resources;
- minimize environmental pollution and degradation; and
- comply with environmental regulations.

Necessities of Sustainable Value Management

It has been explained in the previous sections of this chapter that sustainability issues can be integrated into value management study: this will lead to sustainable value management and help in achieving sustainable construction projects. However, for sustainable value management to be effective and efficient, there are some basic prescriptions that should be followed (Abidin and Pasquire 2005; AlSanad 2015; Gan et al. 2015; Jollands et al. 2015; Roufechael et al. 2015; Ruparathna and Hewage 2015; Saleh and Alalouch 2015; Chang et al. 2016; Al-Yousefi 2016) and these include, among others:

- Sustainable value management should be carried out before 40% completion of the project design stage;
- The objective of the study, which is sustainability and value, should form the motto and goal of the study;
- The Pareto principle of 80/20, indicating that 20% of sustainable value management efforts will give rise to 80% of sustainable construction, should be applied;
- There is a need for a positive contribution of effective and efficient ideas from project team members, especially the designers and project manager;
- The right mix of team members (in term of profession and team roles as discussed in Chap. 3) should be ensured for sustainable value management;
- Whole-life costing of the existing project proposal should be established before the commencement of sustainable value management study;
- The principle of lean construction should be adopted; and
- Other construction management principles and techniques, as discussed in Chap. 7, should be incorporated where necessary for effective and efficient sustainable value management.

Principles of Sustainable Value Management Model

This section examines various principles of a typical value management study that are important in the achievement of sustainable construction goals. Sustainable value management can be attained for projects in the construction industry, provided the following principles are adhered to:

Teamwork

Sustainable value management should be built around teamwork where individuals recognise their importance and that of the other members in the effective and efficient delivery of the goal of the study. Moreover, there should be effective coordination of participants, especially during the speculative phase, while issues relating to disputes and conflicts should be resolved amicably as soon as possible.

Order of Activities

The systematic and logical approach as discussed in Chap. 3 should be adopted for a successful sustainable value management exercise. The job plan structure and framework should be useful for achieving a step-by-step process of sustainable goals. This will ensure that sustainability issues are discussed at various stages for better decision making and policy formulation.

Function Analysis Technique

This is one of the strengths of a typical value management study: this process of VM should be efficiently and effectively executed for sustainable value management. Sustainability issues and goals can be highlighted as part of the functions so that the analysis can be effective and efficient for the selection of the best alternative ideas.

Principle of Unnecessary Cost

The principle is that every element, part or the project itself contains aspects that do not provide any function and therefore constitute unnecessary parts. The cost of providing these parts is known as unnecessary cost. The principles of functional analysis, life cycle costing, brainstorming, and job plan in sustainable value management is to eliminate these costs in order to provide construction projects delivered to value and sustainable goals at the lowest possible cost.

Decision-Making Techniques

The adoption of such tools and techniques as whole-life costing, brainstorming, the Pareto principle, and FAST diagramming will assist in making necessary and appropriate decisions that aid the overall improvement of a construction project. This implies that sustainability issues can be incorporated in a sustainable value management where these techniques are adopted.

Team Dynamic

The involvement of individuals with the ability to contribute to and improve a value management study is one of the principles of a value management study. For sustainable value management, it is necessary to include a sustainability expert or members of regulatory bodies tasked with the responsibility of ensuring sustainable development. Alternatively, it would be preferable if some of the professionals in the VM team have adequate knowledge of sustainable development and what it entails. This will not only be necessary for the experts to make a positive contribution but will also help in the proper understanding of the sustainability goals and how they can be achieved. Various sustainability checks, ratings, and guidelines can also be provided and adopted for analysing the functions of various parts and elements of the project.

Timing of Value Management Study

A sustainable value management study should be carried out during the design phase of preconstruction stage, preferably at the scheme design stage, more preferably at the feasibility or outline proposal stages and less preferably at the detailed design stage. There are various schools of thought on the timing of value management but the idea is that at the scheme design stage, the necessary basic details are already available for the study and the outcome of the study can be incorporated into the next stage of the project detailed design. The detailed design can then be used to prepare production information for tender action and the subsequent award of contract. This helps to ensure that necessary issues of sustainability and other concerns are discussed before final contract decisions are made.

Skill Mix

The facilitator and leader of sustainable value management should ensure that the right set of people in terms of their profession, relevance, contribution, and team roles is included in the team. This will help in harnessing the potential and knowledge of individuals for the performance of the study.

Facilitator

A facilitator has already been described as the leader of the team and for sustainable value management, the facilitator should be experienced in the practice and project area, and possess the necessary skills for managing activities, processes, and other team participants. This will help in creating a better awareness of sustainability issues among the members and will also aid the continuous monitoring of these issues throughout the study stages.

Study Report

Some value management studies end at the presentation phase but sustainable value management should include the important post-study phase as well. This stage involves the preparation of a study report, the monitoring of the implementation of recommendations in the project, and the follow-up. The report will not only help to understand and monitor sustainability issues of the project for which it was prepared, but will also be useful for new projects, especially similar ones, as a starting document for other sustainable value management studies.

Implementation Monitoring

As indicated earlier, this is one of the post-study phase activities of a sustainable value management study. The monitoring of implementation will ensure continuous attention to sustainability issues as well as ensuring proper implementation of the recommendations. The facilitator or the designated member of the sustainable value management team can shed light on grey areas.

Follow Up

This is the last stage of the post-study exercise and it involves data gathering after the completion of the project. In fact, based on the whole-life principle of sustainable value management, follow up is expected to continue to usage, conversion, or re-use and eventually the demolition stage. This will help to monitor the sustainability issues throughout the lifespan of the project. The report can also be used to convince future clients of the importance and benefits of sustainable value management in achieving not only value for money and return on investment, but also the sustainable goals of construction projects.

Stages of Sustainable Value Management

The fundamental principle of sustainable construction is to plan and deliver projects that are affordable, efficient, effective, and provide value to clients and users—present and future—while increasing economic sustainability and decreasing environmental impacts (Bal et al. 2013; Aghimien and Oke 2015; Roufechael et al. 2015). The transformation to sustainable construction has become a necessity not only for universal competitiveness but also for the economic growth of any country (AlSanad 2015; Ametepey et al. 2015; Saleh and Alalouch 2015; Jollands et al. 2015; Mousa 2015; Oke et al. 2015; Ruparathna and Hewage 2015; Yekini et al. 2015; Chang et al. 2016; Kibwami and Tutesigensi 2016; Alwan et al. 2017). For instance, it was posited that all new domestic and commercial buildings in the UK were scheduled to produce zero carbon by 2016 and 2020 respectively.

Value management is usually conducted at the preconstruction stage but issues discussed during the study are not only limited to this stage but concern the entire life of the project. However, modern value management (now referred to as sustainable value management) does not only consider the lifespan of a project from an economic perspective but also environmental and social issues that make the project function as required, and serve the people after construction and for future purpose. This goal is synonymous with sustainable construction which is concerned with projects that cater for economic, social, and environmental issues for the current and future generations.

Sustainability goals are expected to be incorporated into value management at various stages of the study. The sustainability issues that should be discussed and included as part of value management activities are discussed under each of the identified phases of value management (Abidin and Pasquire 2005; Al-Yami and Price 2005; Karunasena et al. 2016; Al-Yousefi 2016). The aim of this section is not to explain the meaning and issues expected at these phases—they have been explained in Chap. 3 of this book – but to shed more light on the sustainability issues. The sustainability issues should be examined and evaluated in conjunction with identified and discussed activities of various value management stages.

Pre-study Phase

At the pre-study phase of sustainable value management, the following activities are expected to be carried out:

Obtain Project Data and Information

All necessary and relevant information regarding the project is to be collected from the clients and design team members. This may include the feasibility report, planning approval, building approval, and statutory regulations.

Define Project Scope, Goal and Requirements

Scope is concerned with the ability and extent to which the project is expected to meet the requirements of the client. It is necessary to understand the main needs of the client, the basic project requirements, and the scope of meeting the objectives. These should be separated from primary needs or requirements and other secondary objectives of the project. The goals should be clearly stated and explained without any form of ambiguity.

Obtain Key Documents

Depending on the stage of introduction and application of value management, the available project documents may vary. However, all the available documents with any relevant information on the project should be obtained at this stage. These documents may include various forms of drawings, specifications, and a breakdown of estimate.

Identify Strategic Issues of Concern

For every project, there are issues of concern in the order of priority. It is necessary for sustainable value management that all issues in terms of challenges and expectations relating to the project are identified. These should be clarified and discussed in detail for proper understanding for the purpose of prioritizing them according to their level of importance and significance to the project.

Determine Scope and Objectives of the Study

It is vital that the extent of the study as well as the expected goals and objectives of sustainable value management are stated in clear terms. Moreover, as discussed in Chap. 7, value management can adopt principles of other construction management techniques such as lean construction, risk management, and total quality management. However, for construction projects where any of the techniques are also adopted, it will be necessary to determine the scope of the sustainable value management to exclude some of the features of the other methods in order to avoid a clash of ideas and recommendations.

Whole-Life Costing of Project and Elements

It is necessary to estimate the overall cost of the project based on the information available from the original project team. This includes initial, annual, running, maintenance, conversion, and demolition costs with a view to forming the basis for the sustainable value management study.

Identify Benchmarks

A benchmark is a standard or basis against which an objective can be evaluated or assessed. At pre-study phase, it is necessary to identify various benchmarks against which the success of sustainable value management will be measured. Various lessons learned from the client and other project stakeholders should also be documented as a guide to the process and practice of the study.

Prepare Models

The existing theories regarding the project should be modified and converted to models that will be easy to understand and interpret by the sustainable value management team members. It is necessary at this stage to prepare models such as cost, quality, energy, and value that will guide the activities and practice of the study.

Gather Appropriate Customer/User Information

At the pre-study phase, it is necessary to collate appropriate information regarding the would-be users or customers of the project. This will help the sustainable value management team to understand the expectations of the project and include them as study objectives in ensuring that the project satisfies the customers and eventual users.

Information Phase

At the point of gathering information for a sustainable value management study, it is necessary to consider the following basics.

Minimization of Resources Consumption

Necessary information should be gathered on how resources consumption can be minimized using the concept of lean construction. Waste is one of the reasons for unnecessary costs and the sustainable value management team must source the necessary statistics, including facts and figures, to ensure this practice.

Identification of Renewable or Recyclable Resources

One of the concerns of sustainability is the ability to convert or re-use resources to avoid redundancy. This implies that materials and resources that have the two attributes are deemed to be sustainable and fundamental to a sustainable value management study. Such items should be identified and details regarding their quality and cost should also be gathered at this stage.

Maximization of Resource Reuse

After identifying the renewable and recyclable resources, sustainable value management should be geared at making use of these as much as possible in preference to non-renewable ones.

Protection of the Natural Environment

The environmental element of sustainability is the oldest and most common concern. To ensure a sustainable value management study, information on the do's and don'ts for protecting the environment should be gathered from relevant sources. Moreover, the focus should be on quality in creating the built environment.

Design for Minimum Waste

Information on lean management should be sought in order to adopt the principles of lean construction for waste reduction and minimization. This should be incorporated into a sustainable value management study. It is possible to engage a lean construction expert for this information.

Conserving Resources

To ensure sustainable value management, information should be sought on the best ways to conserve basic resources for construction projects. These include natural resources such as water, various forms of energy, fuel, and the like.

Consideration of People

This is the social element of sustainability that considers the quality of life of people who are affected by the construction projects now and in the future, as well as in the immediate and distant environment. Information should be gathered on the people who will be affected both negatively and positively, and should form part of the requirements for sustainable value management.

Analysis of Function Phase

Apart from the basic activities expected to be carried out at this stage, as discussed in Chap. 3, this section further explains sustainable value management actions to be executed in relation to sustainability issues.

Identification of High Potential Sustainable Construction Issue

At this stage, it is necessary to identify various sustainability issues that are peculiar to the project under consideration. These should also be classified, categorised, and prioritized based on the level of importance and significance to the client's need, project requirement, statutory regulations, among others.

Identification of Primary Function of Sustainable Component

Like any other function analysis, it is important to identify and evaluate the primary function and purpose of sustainable related activities, elements, or components of the construction project. It is possible the primary function of the component is not related to sustainability but the primary sustainability issue that the component is expected to fulfil should be identified for further action.

Identification of Other Functions of Sustainable Component

This aspect of the study will answer the question 'What else can it do?' in a function analysis. It is important to identify other functions of sustainable components: these are known as secondary functions.

Estimating the Cost of Sustainable Construction Element

After identifying the primary and secondary functions of the sustainable component of the project, it is also necessary to estimate the whole-life cost of providing these functions. These will be evaluated with other alternative ideas in order to select the one with the best value.

Estimation of Value of Sustainable Construction Component

As discussed earlier, value in value management study is related to functions of various components of the construction project as well as the cost of providing them. The function may be evaluated in term of quality, satisfaction derived, and meeting of required need. This should also be carried out for identified sustainable components for an effective and efficient sustainable value management.

Creative Phase

In suggesting ideas during sustainable value management, the following sustainability issues should be considered:

Minimizing Energy in Use

The ideas that should be generated should be in line with the practice of minimizing the use of energy, both for the construction and post-construction phase of the construction projects. The design of the project should be able to conserve energy to the optimum.

Conserving Water Resources

Water is one of the basic natural resources and is used in various ways and forms. As a result of the demand for water, it is one of the natural resources that need to be conserved during construction and during the usage stage of the project. Ideas generated at the creative phase of sustainable value management should therefore be in line with the conservation of water.

Enhance the Quality of Life and Offer Customer Satisfaction

Meeting the needs of the client, who can be customers or end-users, is one of the basic objectives of any project. Ideas at the creative stage of sustainable value management should be geared towards enhancing the quality of life of the people, meeting their expectations and offering them satisfaction.

Provide and Support Desirable Natural and Social Environments

The ideas at the creative stage should be geared towards ensuring that the natural and social environments conform to the health, comfort and aesthetic needs of the people who live there, both now and in the future'. The idea should also reduce the various forms of pollution (air, water) which are harmful to the environment.

Maximize the Efficient Use of Resources

The major resources for construction projects are man, material, machine, and methods. A sustainable value management should take these resources in their various forms into consideration and consider how they can be used effectively and efficiently.

Decrease the Use of Raw Materials

Raw materials are usually sourced directly from the environment except for finished products of some industries that serve as raw materials for others. In construction, the majority of the raw materials come directly from the environment. It is therefore necessary at the speculative phase of sustainable value management to consider generating ideas that will decrease the use of these materials.

Optimize the Consumption of Renewable Resources

In creating ideas for sustainable value management, optimization of the use of renewable and recyclable resources should be ensured. This implies that members of the team should be concerned with the adoption of materials and other resources that will not constitute a threat to the environment by virtue of their redundancy.

Decrease the Amount of Harmful Emissions and Waste

At the speculation phase, it is necessary to adopt materials that are environmentally friendly. In cases where it becomes inevitable, alternatives with the least harmful emissions should be adopted. Moreover, waste generation of the alternatives should form the basis for coming up with ideas.

Enhance Indoor Environmental Quality

One of the social sustainability issues is the enhancement of good indoor environmental quality. As a result, sustainable value management should be concerned with the creation and generation of ideas that will enhance and improve the quality of life of the project users by ensuring that the ideas generated satisfy the indoor environmental quality requirements.

Evaluation Phase

The identified alternative ideas should be evaluated based on the following sustainability issues:

Offer Flexibility and the Potential to Cater for User Changes in the Future

It has been mentioned that sustainability issues are not only designed to cater for the people and environment of the present day, but also include concern for the ability of those to come in the future to be able to enjoy the same benefits. In view of this, one of the factors that should be used for evaluating alternative ideas in sustainable value management should relate to flexibility and the potential to cater for future users and customers.

Maximize the Efficient Use of Resources

Various alternative ideas should be judged by their level of resource usage. Using this as the basis, the best idea at the evaluation phase of sustainable value management is the one that maximizes efficient use of various forms of resources.

Quality Over Entire Project Life

Other factors to consider when judging alternative ideas are their reliability, durability, and serviceability over the life of the project. The best alternative idea should be capable of increasing or maintaining these variables without compromising other aspects of sustainable goals.

Minimizing Non-renewable Energy Consumption

The best idea should maximize the use of recyclable and renewable resources while minimizing the use of non-reusable ones. It is one of the fundamental principles with which the evaluation phase of sustainable value management should be concerned.

Using Environmentally Preferable Products

The alternative idea should be able to adopt products and materials that are friendly to the environment. Injurious or toxic materials should only be used where necessary, provided there are no others that can serve the same function.

Conserving Water

This is concerned with the conservation and minimization of the use of water. The alternative ideas should be evaluated based on their capacity to conserve the use of water at construction and post-construction (usage, conversion, and demolition) phases of the project.

Enhancing Indoor Environmental Quality

This is another criterion for judging and selecting the best alternative ideas. The initiative at this point is that an idea, material, or component that will help to improve indoor environmental quality and improve the quality of life of people is sustainable and preferred over others.

Optimizing Operational and Maintenance Practices

Operational and maintenance costs are some of the important elements of whole-life costing. The selection and use of elements with high maintenance are not economical and as a result, alternative ideas in sustainable value management should be evaluated, not only by their initial cost, but also by their operational and maintenance costs.

Development Phase

To further develop the best alternative ideas in view of a presentation to the body that commissioned the value management study, the following should be ensured:

Select Ideas for Further Action

At the early stage of this phase, the best sustainable alternatives should be highlighted based on the results of the evaluation phase. A written document of the selected ideas should be prepared for further development.

Assess and Refine Sustainability Issues

This is concerned with another round of assessment and evaluation of sustainability issues raised at the pre-study and information phases of sustainable value management. The purpose is to ensure that the selected alternative ideas respond to various sustainability issues.

Conduct Benefits Analysis

The benefits of various proposals, including selected sustainability alternatives, should be identified and explained at this stage of sustainable value management. The benefits may be to clients, customers, users or other concerned stakeholders. Other means of identifying and classifying the benefits can also be adopted.

Estimate Cost Difference of Proposals

This is concerned with the cost analysis of various alternative proposals. The cost should be all inclusive of the initial, final, running, maintenance, conversion, or re-use and demolition costs. Afterwards, it is necessary to prepare a cost-difference analysis of various proposals emanating from the sustainable value management study while giving more attention to the adopted sustainable proposal.

Generate Sketches and Information Needed to Convey the Concept

Drawings of various kinds, bills of quantities, specifications, and other necessary supporting documents that detail the conclusions and recommendations of the sustainable value management study should be generated. The information contained in the documents should be simple, clear, and unambiguous to understand, interpret, and execute by the original project team.

Confirm That a Proposal Should Be Further Developed

Based on the documents containing the recommendations of value management exercise and other relevant information, it is necessary that sustainable value management study accomplish its set objective. If this is not the case, it may be necessary to further develop the proposal until the set goals at the pre-study phase are accomplished.

Presentation Phase

The activities at this stage are purely value management activities and it is expected that all issues regarding sustainability would have been incorporated for presentation. The following activities are expected at this stage:

Prepare Presentation and Supporting Documentation

The first activity at this stage is to prepare documents including drawings, sketches, maps, PowerPoint slides and other necessary supporting documents that will aid a proper and professional presentation of the sustainable value management report.

Outline an Anticipated Implementation Schedule

The sustainable value management team should also prepare various means of implementing the recommendations emanating from the study. An action plan and the construction stages of implementing them should also be stated and explained.

Exchange Information with Project Team

Another important activity of the presentation phase is the exchange of information with the design team. At this stage, the sustainable value management team, through the facilitator or designated member(s), should ensure that the project team understands the recommendations and how they will be applied to the project for which it was conducted. It may also be necessary to discuss various alternative ideas and the criteria adopted for selecting the best alternative ones.

Ensure Management Has Full Information

The management team of the construction project, especially the project manager or lead consultant, should be adequately carried along at the presentation phase. This is to ensure that stakeholders in charge of the project have the necessary and detailed objective information for making the best decisions.

Also at the presentation stage, the two teams, namely the project team and the sustainable value management team, should meet, not just to discuss the results of the study, but also to examine how the recommendations will be implemented.

Post-workshop Phase

This stage entails a review of the study, the implementation of the sustainable value management recommendations, and the follow-up of construction project activities. To ensure sustainable construction, it is necessary at this stage to conduct implementation meetings, establish action plans for the accepted alternatives, obtain commitment for implementation from project team members, and track value achieved due to the implementation of the best alternatives.

The itemised and discussed activities are not exhaustive: it the duty of sustainable value management team members, especially the facilitator, to ensure that the necessary activities are included and executed in the study.

Summary

Adopting sustainability concepts in value management practice is termed sustainable value management and this chapter explained various issues and necessities for the practice. The chapter discussed various factors to be considered before initiating a sustainable value management study. It also explained the suitability of the study for construction projects by discussing various forms of projects that will benefit significantly from its adoption. Moreover, a sustainable value management model incorporating cost, value, quality, and time around flexible sustainable goals was discussed and the model was related to three elements of sustainability, namely, economic, social, and environmental. Further to this, the necessities, principles, and stages of sustainable value management study were identified and explained accordingly.

References

Abidin, N. Z., & Pasquire, C. L. (2005). Delivering sustainability through value management: Concept and performance overview. *Engineering, Construction and Architectural Management, 2*(2), 168–180.

Aghimien, D. O., & Oke, A. E. (2015). Application of value management to selected construction projects in Nigeria. *Developing Country Studies, 5*(17), 8–14.

AlSanad, S. (2015). Awareness, drivers, actions, and barriers of sustainable construction in Kuwait. *Procedia Engineering, 118*(2015), 969–983.

Alwan, Z., Jones, P., & Holgate, P. (2017). Strategic sustainable development in the UK construction industry, through the framework for strategic sustainable development, using building information modelling. *Journal of Cleaner Production, 140*(2017), 349–358.

Al-Yami, A. M., & Price, A. D. (2005). Exploring conceptual linkages between value engineering and sustainable construction. In F. Khosrowshahi (Ed.), *21st Annual ARCOM Conference* (pp. 375–384). Association of Researchers in Construction Management: University of London.

Al-Yousefi, A. S. (2016). The synergy between value engineering and sustainability. *Presentation at the 8th World Congress of the Council on Tall Buildings and Urban Habitat (CTBUH)*. Dubai. Retrieved November 14, 2016, from http://www.energyandwateroman.com.

Ametepey, O., Aigbavboa, C., & Ansah, K. (2015). Barriers to successful implementation of sustainable construction in the Ghanaian construction industry. *Procedia Manufacturing, 3* (2015), 1682–1689.

Atkinson, R. (1999). Project managemement: Cost, time and quality, two best guesses and a phenomenon, its time to accept other success criteria. *Journal of Project Management, 17*(6), 337–342.

Bal, M., Bryde, D., Fearon, D., & Ochieng, E. (2013). Stakeholder engagement: Achieving sustainability in the construction sector. *Sustainability, 6*(5), 695–710.

Chan, A. P. (2001). Framework for measuring success of construction projects. Brisbane. Retrieved November 27, 2016, from www.construction-innovation.info.

Chang, R., Soebarto, V., Zhao, Z., & Zillante, G. (2016). Facilitating the transition to sustainable construction: China's policies. *Journal of Cleaner Production, 131*(2016), 534–544.

Gan, X., Zuo, J., Ye, K., Skitmore, M., & Xiong, B. (2015). Why sustainable construction? Why not? *An owner's perspective. Habitat International, 47*(2014), 61–68.

Jollands, S., Akroyd, C., & Sawabe, N. (2015). Core values as a management control in the construction of 'sustainable development'. *Qualitative Research in Accounting & Management, 12*(2), 127–152.

Karunasena, G., Rathnayake, R. M., & Senarathne, D. (2016). Integrating sustainability concepts and value planning for sustainable construction. *Built Environment Project and Asset Management, 6*(2), 125–138.

Kibwami, N., & Tutesigensi, A. (2016). Enhancing sustainable construction in the building sector in Uganda. *Habitat International, 57*(2016), 64–73.

Mousa, A. (2015). A Business approach for transformation to sustainable construction: An implementation on a developing country. *Resources, Conservation and Recycling, 101*(2015), 9–19.

Noor, N. F., Kamruzzaman, S. N., & Ghaffar, N. (2015). Sustainability concern in value management: A study on government's building project. *International Journal of Current Research and Academic Review, 2*(2015), 72–83.

Oke, A. E., Aghimien, D. O., & Olatunji, S. O. (2015). Implementation of value management as an economic sustainability tool for building construction in Nigeria. *International Journal of Managing Value and Supply Chains, 6*(4), 55–64.

Roufechael, K. M., Abu Bakar, A. H., & Tabassi, A. A. (2015). Value management and client attitude in developing sustainable construction. *Advances in Environmental Biology, 9*(5), 4–6.

Ruparathna, R., & Hewage, K. (2015). Sustainable procurement in the Canadian construction industry: current practices, drivers and opportunities. *Journal of Cleaner Production, 109* (2015), 305–314.

Saleh, M. S., & Alalouch, C. (2015). Towards sustainable construction in Oman: Challenges & opportunities. *Procedia Engineering, 118*(2015), 177–184.

Shahu, R., Pundir, A. K., & Ganapathy, L. (2012). An empirical study on flexibility: A critical uccess factor of construction projects. *Global Journal of Flexible Systems Management, 13*(3), 123–128.

Shen, G. Q., & Yu, A. T. (2012). Value management: Recent developments and way forward. *Construction Innovation, 12*(3), 264–271.

Tabish, S. Z., & Jha, K. N. (2011). Identification and evaluation of success factors for public construction projects. *Construction Management and Economics, 29*(2011), 809–823.

Takim, R., & Akintoye, A. (2002). Performance indicators for successful construction project performance. In D. Greenwood (Ed.), *18th Annual ARCOM Conference 2* (pp. 545–555). Northumbria: Association of Researchers in Construction Management.

United Kingdom. Cabinet Office. (2011). *Government construction strategy.* London: Cabinet Office. Retrieved November 14, 2016, from http://www.gov.uk.

Yekini, A. A., Bello, S. K. & Olaiya, K. A. (2015). Application of value engineering techniques in sustainable product and service design. *Science and Engineering Perspectives, 10*(Sept), 120–130.

Chapter 9
Stakeholders to Sustainable Value Management

Abstract This chapter explains how project stakeholders and participants of sustainable construction can be incorporated into value management practice and study in the quest of achieving sustainable construction projects. Stakeholders to sustainable value management team were identified as internal or external members. Cost of value management, which is related to the selection and number of team members, was noted as one of the risks of value manager. The size of the team should be large enough to allow for the generation of various ideas, brainstorming and delegation of responsibilities; it should also be few enough to reduce the cost of the study. Care should therefore be taken by the facilitator of sustainable value management to ensure that there are not too many participants when trying to include all relevant members in the team.

Keywords Construction project · Stakeholder management · Sustainable value management · Team composition · Value management · Value management facilitator

Introduction

Having explained value management as a sustainable construction tool in Chap. 8 and having discussed the importance of project stakeholders and participants to sustainable construction, this chapter explains how these individuals can be incorporated into value management practice and study in achieving sustainable construction projects.

As discussed in Chap. 3, there are various stakeholders of diverse interests to construction projects (Olatunji et al. 2014; Nash et al. 2010; Zanjirchi and Moradi 2012; Bal et al. 2013; Aapaoja and Haapasalo 2014). However, this chapter explains their roles and responsibilities to sustainable value management, whether as internal or external members of the team. Internal members in this case are

A.E. Oke and C.O. Aigbavboa, *Sustainable Value Management for Construction Projects*, DOI 10.1007/978-3-319-54151-8_9

individuals selected by the facilitator to join the sustainable value management team while external members are not part of the team but can influence the initiation, outcome, and adoption of recommendations of the team.

Facilitator

The facilitator is the leader of the value management team and is tasked with the responsibility of planning, controlling, and coordinating the study. In view of this, anyone assigned to the position of facilitator of sustainable value management for construction projects is expected to be knowledgeable about the basics of the practice, understand how it works, and have the necessary experience of construction activities and processes, and the knowledge of materials science. It is possible that other members of the team are not value managers by training, experience or practice but the facilitator, who is an internal member, should have these basic skills, competencies, and training to function effectively and efficiently as the team leader.

Required Skills for Facilitator

In order to facilitate sustainable value management successfully, there are some basic personal skills attributes expected of a facilitator (Daddow and Skitmore 2005). Some of these include the following:

- Smart thinking;
- Approaching issues objectively;
- Having an open mind to communication;
- Having an inquiring mind;
- Being attentive;
- Being adaptable and flexible;
- Having confidence;
- Having lateral thinking ability and intuition;
- Being proactive;
- Knowing the project requirements;
- Being motivated and enthusiastic;
- Adopting a positive and constructive approach to issues;
- Having personal skills;
- Having an open mind towards ideas and people;
- Communicating ideas confidently and professionally;
- Listening to ideas of others;
- Relating with others;
- Bringing expertise to the sustainable value management study;

- Having expertise and experience of the construction industry;
- Recognising reactions, whether verbal or physical; and
- Understating people's statements and what they mean.

Expected Competencies of Facilitator

There are required competency areas for value managers (Sigle et al. 2000; Oke and Ogunsemi 2013). These include the following:

- Creativity in developing new ideas and solutions;
- Mental alertness to responsively react to ideas, suggestions and innovations;
- Transformational leadership;
- Listening skills;
- Collaborative conflict management through negotiation and cooperation;
- Expressive and interactive social style;
- Innovation;
- Self-motivation; and
- Abstract reasoning and logical thinking through imaginative ideas generation.

Required Training Areas for Facilitator

A facilitator of sustainable value management is required to have basic training in value, cost, function, sustainable goals, and other important areas (Leung 2001; Michel 2001; Oke and Ogunsemi 2012). The following are some of the training requirements for a facilitator:

- Training on value management history and concept;
- Training on value management approaches and techniques;
- Training on value management workshops for construction projects;
- Training on function analysis;
- Training on cost analyses;
- Training on job plan and workshop ethics;
- Training on value in relation to sustainable construction project with the emphasis on social, economic, environment, psychological;
- Training on decision analysis and techniques;
- Training related to project management;
- Training on personal skill attributes required of value manager;
- Training on communication and presentation skills;
- Training on conflict resolution;

- Training on leadership and other people management skills;
- Training on qualitative and quantitative investigative methods; and
- Training on report writing.

Quantity Surveyor (Cost Estimator)

This is an important member of a sustainable value management team owing to the necessity of the team's preparing estimate and cost of alternative ideas. The quantity surveyor is also known as an estimator, cost engineer, and construction economist. The experience of a quantity surveyor in preparing a cost analysis for proposals arising from the study is vital to the success of sustainable value management. The quantity surveyor should be an internal member and his or her duties will include the following:

- Provide necessary cost-related information;
- Source for further cost and associated information for the team;
- Prepare whole-life costing of original design;
- Provide cost-related information of alternative ideas;
- Provide information on cost-related factors such as inflation and the exchange rate;
- Prepare estimate for alternative ideas;
- Prepare estimate for alternative designs;
- Prepare investment appraisal for proposals;
- Prepare whole-life costing of best alternative design;
- Advise on legal and other construction issues; and
- Perform other team-related functions.

Architect

The presence of an architect in a sustainable value management team may depend on the type of project but they are required for all forms of construction projects in some countries. This implies that an architect can be an internal or external member of the team. Also known as the designer, an architect in a sustainable value management team is expected to perform the following functions:

- Provide information on design-related issues for the team;
- Source for further information on design-related issues;
- Examine original project design for sustainability and other considerations;
- Prepare design for various alternative ideas;
- Provide information on buildability of the ideas;
- Prepare design for alternative proposals;

- Prepare detail drawings for the best alternative proposal;
- Advise on legal and other construction issues;
- Be involved at post-study phase for implementation of sustainable value management recommendations by original project team and advise, especially when there are issues with design drawings; and
- Perform other team-related functions.

Engineer

An engineer of a particular discipline can be engaged in sustainable value management study, depending on the nature of the project. For instance, it will be ideal to involve a civil or structural engineer for civil engineering projects for informed data that can help the study. As an internal member of a sustainable value management team, the responsibilities of an engineer may include the following:

- Provide information on engineering design-related issues for the team;
- Source for further information on engineering design-related issues for the team;
- Examine original project design for sustainability and other considerations;
- Prepare engineering design for various alternative ideas;
- Provide information on buildability of the ideas;
- Prepare engineering design for alternative proposals;
- Be involved at post-study phase for implementation of sustainable value management recommendations by original project team and advise, especially when there are issues of an engineering nature;
- Prepare detail engineering drawings for the best alternative proposal;
- Advise on legal and other construction issues; and
- Perform other team-related functions.

Project Manager

The project manager can be an internal or external member of the sustainable value management team. As a member of the team, the construction project manager can contribute to the success of the study by providing administration details required by members, helping in the coordination of team goals and members, as well as performing other duties designated to him or her. A project manager can also initiate and conduct sustainable value management as part of his or her responsibilities for the project. In that case the project manager can function as the facilitator for the study—provided that he or she possess the experience, training, and professionalism—or use the recommendations from the study to make decisions during the project execution.

Construction Manager

The construction manager is also known as the builder. The role of a construction manager is vital to the success of construction project and it will be of great value if also included in a sustainable value management team. As a member of the team, the following will be required of a construction manager:

- Provide information on building production management;
- Provide information on buildability;
- Provide information on maintenance and other issues on alternative ideas;
- Be involved at post-study phase for implementation of sustainable value management recommendations by original project team;
- Advise on legal and other construction issues where necessary; and
- Perform other team-related functions.

Other Construction Team Members and Professionals

Depending on needs of the client, user requirements, the type of project, and other associated features, it may be necessary to involve some other construction professionals as internal members of the value management team. These may include a land surveyor, estate surveyor, and town planner among others. For instance, a town planner may be required as an internal member of the sustainable value management team for opening up, planning, and providing basic infrastructure for new community development. It is therefore the responsibility of the facilitator to identify the right set and mix of construction professionals for an effective and productive study.

Contractors

Depending on the procurement arrangement, tendering approach, stage of involvement of the contractor as well as the type and nature of the project, a contracting organisation also known as contractor in some cases, may be involved as an internal member of the sustainable value management team. However, such an individual representing the organisation should have the basic skills and experience to contribute positively by generating innovative ideas and providing other necessary information that will be useful for a successful study. In most cases, contractors are external to the study but their importance in interpreting and actualising recommendations from the study is fundamental to the overall success of sustainable value management. The constructing organisation should possess the following in order to execute recommendations emanating from the study effectively and efficiently:

- An understanding of the recommendations from the sustainable value management team;
- Professionals with experience and adequate training to interpret and implement recommendations;
- Workforce with required skills and expertise to implement recommendations;
- Plants and equipment to implement recommendations;
- An ability to source for required materials to implement recommendations; and
- An understanding of the methods, processes, and techniques to implement recommendations.

Clients and Owners

Most value management are commissioned by clients or client organisations acting on behalf of the clients or owners. They are usually external parties to sustainable value management but their role and effect cannot be overemphasized. The activities of the client or owner that may influence sustainable value management include the following:

- Willingness to adopt sustainable value management due to a lack of awareness and experience of the study;
- Willingness to pay for sustainable value management study due to the perception of adding to the original project cost; and
- Willingness to accept recommendations from sustainable value management by interfering and dictating what to accept and reject.

Sponsors or Financiers

For construction projects where the clients or owners are not directly responsible for financing them, the role of the sponsors or financiers becomes unique and important. Depending on the level of agreement between them, the influence of the financier as an external member may impact the adoption of sustainable value management as well as the implementation of their recommendations for construction projects. Except where the cost of value management has been billed into the project cost, the financier may not be willing to pay for the study. However, in some cases, the financing organisations with experience of the practice of value management and their benefits are now advising and insisting that a value management study be carried out for construction projects, especially those that are complex, expensive, and public-related.

Government

Governments are the major clients in any country and are always the pioneers in the application of new concepts, practices, procedures, and techniques. For construction projects, the various arms of government are vital in the promotion and adoption of sustainable value management in the construction industry as observed from countries where the practice has been fully adopted. As a result, value management bodies have been constituted and backed by necessary laws and regulations to regulate their practice. In some countries, the use of value management has been mandated for specific types of projects and many corporate and private clients have followed suit. Learning from these countries, the following are required of countries where the practice is still in its infancy:

- Creating awareness of the practice among related government members by construction professionals' bodies and their regulatory boards;
- Political will and commitment of government to adopt sustainable value management;
- Enactment of regulations and laws to support the adoption and enforcement of sustainable value management;
- Provision of guidance notes and checks for sustainable value management; and
- Enforcement of the regulations by agencies, parastatals, and statutory bodies of the government.

Government agencies, parastatals, statutory bodies and the like who are responsible for the planning, monitoring and regulating of construction projects are key external stakeholders to sustainable value management practice in the construction industry. Right from the inception of the project, their involvement in obtaining planning approval in line with regulations and standards is vital for sustainable construction. Their activities also continue throughout the entire life of the project. In fact, one of the items of information required at the pre-study phase of sustainable value management is the issue of project approval and associated regulations relevant to the project and including the site of construction. The roles of these agencies include the following:

- Checking of suitability of site for construction project;
- Checking of environmental conditions and type of project designated for the area based on the area master plan;
- Checking of drawings' conformity with regulations such as boundary;
- Checking of project location in relation to community basic amenities and services;
- Evaluating of project documents in line with sustainability goals and other regulations;
- Approving initial drawings provided they comply with standards and regulations;
- Providing terms of approval and guidelines for actual project construction;
- Monitoring of project activities in line with term of approval;

- Clarifying conditions and terms where misinterpreted or abused by project team members;
- Ensuring adherence to standards during construction phase of project;
- Ensuring adherence to standards and regulations during usage or operation stage;
- Ensuring adherence to standards during conversion or re-use stage; and
- Ensuring adherence to regulations and standards during demolition of project

End-Users

Concern for end-users is one of the fundamental principles of sustainability and in construction projects it is important to understand their requirements in order to provide projects that will meet their satisfaction. One of the goals of sustainable construction is concern for this set of people. It is therefore necessary that their opinions and requirements are collected and tabled before the sustainable value management team in order to arrive at the best alternative ideas that will satisfy them and the future generations.

Community

People living in the community where the project is to be situated are important stakeholders who should be given priority. Especially in the case of projects that divide public opinion, it may be necessary to include a leader or representative of the community in the sustainable value management team so as to involve them in the generation of sustainable ideas that will be beneficial, not only to the team and project, but the people and society at large.

Sustainability Experts

Members from agencies or bodies tasked with the responsibility of planning, monitoring, and managing sustainable goals in the areas can be involved in sustainable value management, either as internal or external members. It is not compulsory to involve them as internal members but information regarding sustainability in construction can be obtained from this group of people. This will serve as a basis for creating and evaluating alternative ideas. Sustainability experts may include an environmentalist, socialist, economist, lean managers, resource managers, and the like.

Other Stakeholders

Apart from the stakeholders highlighted and discussed above, there are some others who may contribute to a sustainable value management team as internal or external members. Other stakeholders may include suppliers, economists, conservationists, archaeologists, foresters, interest groups, journalists, solicitors and lawyers. Their participation in, contribution to, and influence on sustainable value management may depend on the following:

- Type of construction project (building, civil, industrial or a combination);
- Nature of project in term of complexity, cost;
- Specific project requirement such as a beautiful environment;
- Type of client influencing the choice of stakeholders;
- Type of facilitator and mode of operation;
- Level of influence of stakeholder on project; and
- Regulations necessitating the involvement of specific stakeholder.

Team Roles of Sustainable Value Management Team

Value management is a team-based study that does not only rely on the professional qualification or experience of team members but also on their teamwork abilities. As a result of the involvement of individuals in a sustainable value management team, an important issue in managing and fostering peace in a team is to ensure that individuals with appropriate team roles are selected. There are nine known and well researched team roles that are required for team success. The team roles as well as their requirements at different stages of development were discussed in detail in Chap. 4 of this book under 'team roles of project stakeholders' section. However, this section examines the importance of the team roles to sustainable value management study.

Plant

A plant pays special attention to major issues and is always willing to shoulder responsibilities that look difficult to achieve. In a sustainable value management team, such individuals are needed not only to express themselves without restraint which is required at the speculative stage of the study, but to also show creativity, inventiveness, and imaginative skills when required for the team.

Specialist

Specialists who are skilful in key areas of sustainable value management such as design, costing, and market survey are required for the success of the study. These roles are played most especially by construction professionals trained in the aspect but it would not be surprising if other members of the team are endowed by reason of extensive experience in the construction field with talent attributed to a profession. A specialist is always focused on the project at hand and as a result of his or her specialisation, a minimum amount of time is spent on achieving a task and as such, helps in guarding against waste of time and effort which is important to a value management study.

Completer or Finisher

Finishing an exercise or activity is fundamental to sustainable value management as there are number of these to be completed within the agreed time set for the study. A completer or finisher ensures that mistakes are avoided by trying to avoid errors due to omission and commission, thereby ensuring that tasks are completed to schedule, standard, and cost.

Implementer

At each phase of a value management study, especially the creative phase, it is expected that various ideas will be generated before they are evaluated and developed to sensible ones. The role of an implementer is to ensure that every idea, concept, view, and opinion is followed through to a practical working procedure. Such a person is methodological and follows procedures through. This is a basic principle of a sustainable value management study and the importance of this individual to the study cannot be overemphasized.

Team Worker

A team worker is interested in the togetherness of a team and is ready to forego his or her opinion for that of others, solely for team cohesiveness and progress. In a sustainable value management study, as in other team activities, there are bound to be misunderstandings and some members with low self-esteem. A team worker is thus needed to support the ideas of others, even when their suggestions or concepts look odd, thereby help in improving communication and fostering the team spirit of members.

Monitor Evaluator

To monitor is to ensure proper conduct by checking for incorrect or unfair behaviour, activities, and the like. A monitor evaluator in a sustainable value management team performs these tasks as well as an examination of ideas and concepts arising from the team. This is based on value-associated features such as function, quality, condition, extent, and importance. Such a person is impartial and trustworthy and can be relied on to resolve disputes or conflict that may arise in the discharge of duties of the team.

Shaper

A shaper mostly directs the effort, attention, and priorities of the team in the right manner, regardless of the situation or condition of the team and team members. Such a person may be domineering and impatient with the sole objective of ensuring that things are done in a proper way. In sustainable value management, people of diverse behaviours, concerns, attitudes, and dispositions are involved and such an individual with shaper attributes is required to keep proceedings in order and put things in the right shape.

Co-ordinator

A co-ordinator is also known as the chairman and his or her responsibility is to organise and manage the team to achieve their set goals and objectives. In a sustainable value management study, it is expected that the facilitator who is the leader of the team will possess this attribute, either at primary or secondary level, to ensure that other members of the team focus on the same principle of generating ideas for the improvement of the function of elements, components, parts or the entirety of the project. However, it is not out of order if some other members of the team possess the virtues of a co-ordinator by trusting others, being stable in their disposition and always being positive about individuals and team success.

Resource Investigator

Sourcing for the right and relevant information is a major principle upon which the sustainable value management process thrives, especially at the information, creation, and evaluation phase of the study. It is required that some of the members exhibit resource investigator skills by sourcing information concerning generated ideas. Such people are expected to be curious, inquisitive, and optimistic in discharging their responsibilities.

Summary

In this chapter, stakeholders to a sustainable value management team have been identified as internal or external members. Cost of value management which is related to the selection and number of team members was noted as one of the fundamental areas of value management exercise in Chap. 2. There should be enough team members to allow for the generation of various ideas, brainstorming, and the delegation of responsibilities; they should also be few enough to reduce the cost of the study. Care should therefore be taken by the facilitator of sustainable value management to ensure that there are not too many participants when trying to include all relevant members in the team.

References

Aapaoja, A., & Haapasalo, H. (2014). A framework for stakeholder identification and classification in construction projects. *Open Journal of Business and Management, 2*(2014), 43–55.

Bal, M., Bryde, D., Fearon, D., & Ochieng, E. (2013). Stakeholder engagement: Achieving sustainability in the construction sector. *Sustainability, 6*(5), 695–710.

Daddow, T., & Skitmore, M. (2005). Value management in practice: An interview survey. *Australian Journal of Construction Economics and Building, 4*(2), 11–18.

Leung, M. (2001). *Educational framework for MSc in value management.* Paper presented at a Conference on towards a Value Management and Value Engineering Educational Framework at Master's Level, Florida.

Michel, J. (2001*). What value societies, value association and value institutes need to see before they give accreditation?* Paper presented at a Conference on towards a Value Management and Value Engineering Educational Framework at Master's Level, Florida.

Nash, S., Chinyio, E., Gameson, R., & Suresh, S. (2010). The dynamism of stakeholders' power in construction projects. In C. Egbu (Ed.), *26th Annual ARCOM Conference* (pp. 471–480). Leeds: Association of Researchers in Construction Management.

Oke, A. E., & Ogunsemi, D. R. (2012). Training of quantity surveyors for value management practice. *International Journal of Construction Project Management, 4*(2), 147–157.

Oke, A. E., & Ogunsemi, D. R. (2013). Key competencies of value managers in Lagos state, Nigeria. In S. Larry & S. Agyepong (Eds.), *Proceeding of 5th West Africa Built Environment Research (WABER) conference* (pp. 773–778). Ghana: Accra.

Olatunji, S. O., Oke, A. E., & Owoeye, L. C. (2014). Factors affecting performance of construction professionals in Nigeria. *International Journal of Engineering and Advanced Technology, 3*(6), 76–84.

Sigle, H. M., Klopper, C. H. & Visser, R. N. (2000). The South African quantity surveyor and value management. *Project Pro*, 23–26.

Zanjirchi, S. M., & Moradi, M. (2012). Construction project success analysis from stakeholders' theory perspective. *African Journal of Business Management, 6*(15), 5218–5225.

Part V
Enhancing Sustainable Value Management in Construction

Chapter 10
Barriers of Sustainable Value Management

Abstract There are several benefits associated with sustainable value management if properly practised and adopted. As beneficial as the study can be to construction projects, there are some barriers that may hinder its adoption, the conducting of the study or the implementation of recommendations emanating from the study. This chapter discusses various limitations and hindrances to the successful initiation and implementation of sustainable value management. Three stages of the barriers to sustainable value management study in construction are discussed. The first is the willingness to accept and adopt the study for construction projects by the initial project stakeholders, especially the clients and lead consultants. This will depend on their experience of the study and their willingness to try something new. The other set of barriers arises during the conduct of the study which may include interference from the project team, and the inappropriate composition of the team. After the study, another barrier to sustainable value management is the willingness to accept and adopt recommendations emanating from the study for the construction projects under consideration.

Keywords Construction project · Stakeholder management · Sustainable construction · Sustainable value management · Team management · Value management

Introduction

There are several benefits associated with sustainable value management if properly practised and adopted. As beneficial as the study can be to construction projects, there are some barriers that may hinder its adoption, the conducting of the study or the implementation of recommendations emanating from the study.

There are three stages of the barriers to sustainable value management study in construction. The first is the willingness to accept and adopt the study for construction projects by the initial project stakeholders, especially the clients and lead consultants. This will depend on their experience of the study and their willingness

A.E. Oke and C.O. Aigbavboa, *Sustainable Value Management for Construction Projects*, DOI 10.1007/978-3-319-54151-8_10

to try something new. The other barrier arises during the conducting of the study which may include interference by the project team, inappropriate composition of the team, etc. After the study, another barrier to sustainable value management is the willingness to accept and adopt recommendations emanating from the study for the construction projects in consideration.

The limitations to the acceptance, usage and implementation of the study are explained in this chapter (Shen and Liu 2004; Abidin and Pasquire 2005; Al-Yami and Price 2005; Daddow and Skitmore 2005; Liu and Shen 2005; Bowen et al. 2010; Bowen et al. 2010; Oke and Ogunsemi 2011; Perera et al. 2011; Leung and Yu 2014; AlSanad 2015; Ametepey et al. 2015; Jay and Bowen 2015; Oke et al. 2015; Roufechael et al. 2015; Yekini et al. 2015; Al-Yousefi 2016; Karunasena et al. 2016). The barriers are classified into three categories, namely study-related, stakeholders and project-related factors.

Barriers Related to the Study

These are the sets of limitations and constraints due to the participant, process, and style of conducting a sustainable value management study. They are related to the timing, cost, and approach of the study as well the knowledge of team members in the two areas of sustainability and value management.

Timing of Value Management

One of the contentious issues regarding value management as a study is the best time for its introduction. As discussed in Chaps. 3 and 8, sustainable value management should be introduced at the preconstruction phase of construction projects before the final designs are prepared for tendering and project planning. Any attempt to introduce the study earlier may not bear fruit as the basic information required for the study may not be available and this will reduce the effectiveness and efficiency of the study outcome. Moreover, the introduction of the discipline later than the suggested phase of project may also be problematic for the project. Adoption of the study after design completion will require that consultants prepare all the drawings all over again if the recommendations from the study are to be incorporated. It can also mean that basis on which the tenders were received from contractors will have to change, which does not represent a good image of the consultants and the project itself. The problems of reconstruction, re-work, cost overrun, and time overrun are associated with implementing the study during construction which is not advisable, except for a mini-study geared towards a particular aspect of the project.

Approach to Value Management

There are several approaches to the facilitation of a sustainable value management workshop and study as discussed in Chap. 3. The earliest and most common one is the 40-h workshop. However, owing to the type of participants and the nature of the project, the selection of an inappropriate method may affect the study negatively. The facilitator should therefore ensure that the best approach is adopted in order to ensure that the study achieves the objectives for which it was commissioned.

Knowledge of Value Management and Sustainable Construction

A major issue of sustainable value management is the team members' knowledge of value management and sustainability goals. It is expected of the facilitator that he or she has the necessary value management experience, training, and skills to manage the team: however, the concern is related to the sustainability aspect. Moreover, the remaining members of the team may not have participated in any previous value management and it is also possible that some or all of them are not knowledgeable regarding sustainable construction. Regardless of the scenario, the likelihood of the study to achieve the desired results may be in jeopardy if not properly managed. It is therefore necessary for the facilitator to ascertain the nature of the experience and training of team members. Experts in value management and sustainability goals can be include or invited while members can also be encouraged to seek further knowledge in the areas before the actual commencement of the study with the view to ensuring an effective and efficient study.

Cost of Value Management

The cost of conducting sustainable value management as discussed in Chap. 2 includes the cost of the venue, participants' fees, and miscellaneous fees, which are considered by some project stakeholders as unnecessary additional costs incurred by client of the project. However, the study needs to be funded adequately for various activities, including information gathering, market survey, and the preparation of documents for various alternatives and proposals. In as much as it has been proved that the cost of conducting a typical sustainable value management study is always a small percentage of the savings derived from the project as a result of implementing the study recommendations, the facilitator should keep the cost of the study as low as possible without compromising the required skills and activities to achieve the optimum output and recommendations.

Team Member Composition for Value Management Exercise

The issue of sustainability involves the consideration of various stakeholders who impact project activities and are impacted by the projects. This makes the issue of sustainable stakeholders a complex one because it involves individuals, organisations, agencies, and bodies of various forms of experience, training, and project interests. For sustainable value management to be effective and efficient, the team members should be carefully selected to include relevant participants who will not only contribute positively to the study objective but also help in convincing other members of sustainability of the recommended ideas. The other members will include project team members, clients, financiers, community, users, regulatory agencies, and the like. The facilitator as the leader of the team should also guide against the imposition of members who will not contribute to the study, thus increasing the unnecessary cost of the study.

Stakeholders Barriers

Stakeholders in this concept refer to external members of a sustainable value management study who have influence on the adoption and implementation of recommendations emanating from the study.

Client Commitment

The commitment of the client to a sustainable value management study will depend on his or her experience of the study, the previous relationship with the facilitator or individual suggesting the study, the willingness of the client to try something new if the practice is new to him or her, and advice from project team members or lead consultant. Although value management has been mandated for some projects in some countries, the client commitment to the study is also important. It is possible that the study is carried out for projects as required by regulatory agencies but their recommendations may be partly adopted or not at all. The facilitator should therefore consider the type of clients and convince them of the advantages of the study to gain their confidence and commitment to the study.

Willingness of Project Team

In most cases, client representatives, also known as consultants, and the project manager or lead consultant in particular, are the major advisers to the project client and their advice is usually what the client acts upon. The team members who are unaware of the study or have a wrong perception of the study may not be willing to support the adoption of the study for projects in which they are involved. There are two major issues with this: one is that some of them are unwilling to change from their current practice in the belief that it has been working over the years and there is no need to fix what is not broken. Moreover, some of the consultants, especially designers and other associated professionals, perceive the study as a threat to their professions since recommendations from the study will have to be incorporated into their original designs. Some of them believe that it is a proof of their incompetence since they have been paid to advise clients in the best possible areas and their job is under threat if such activities continue. However, the facilitator should be able to explain the concept of the study that it is not to replace their service but rather to enhance it by providing information that will collectively help the successful delivery of construction projects.

Wrong Perception of the Exercise

Stakeholders, including construction professionals, have varying views and opinions regarding value management. Owing to a lack of adequate knowledge and training of the practice, some have perceived it as another cost management exercise, a standardisation practice, a conflict-oriented review of design, or a cost-cutting exercise. These barriers of wrong perception can inhibit the adoption of the practice as well as acceptance of recommendations emanating from the study.

Cost management is strictly concerned with the planning, evaluation, monitoring, and controlling of cost throughout the project life cycle, unlike sustainable value management that includes these activities but with the emphasis on value, scope, function, objectives, and standards that the project is expected to meet. The study serves as a standardisation exercise but that is just one of the principles. Using the principle of teamwork of relevant stakeholders, the study reviews the original design, not for conflict resolution but to improve various objectives and functions of the project at the lowest possible cost. Moreover, sustainable value management is not just about cutting costs but the primary focus is achieving a function, using the best alternative material that can be used to achieve the same or better function at the least overall cost. This implies that it is sometimes possible that the best alternative idea or material may be more expensive than that proposed in the original design.

Regulatory Related Issues

In countries where sustainability and value management have not been fully adopted, the major issue has been the lack of policies and regulations to support the practice. There are neither laws to compel practice nor guidelines provided on how to implement and employ them. This is a major hurdle to a sustainable value management study.

Project-Related Barriers

Project related barriers are limitations to sustainable value management that are the direct result of project issues. These include some specific attributes of the project, the adopted procurement method, and the tendering technique, amongst others.

Project Characteristics

The basic attributes or characteristics of construction projects are explained in Chap. 4. In most cases, these types of projects are expensive, complex, risky, complicated, dynamic, multidisciplinary, multifaceted, and unpredictable. Depending on the type and nature of project, the identified attributes can be a major barrier to the adoption of sustainable value management. However, these can also be a positive factor for driving the adoption and application of the study. For instance, projects that are less expensive may not be suitable for the study as the cost of the study may even be more than the cost of the project. Moreover, the confidential nature of some projects is another hindering factor as the client will be concerned that only a few people know about the concept, design, and structure of the project.

Project Procurement Option

The choice of the procurement method is another barrier to the adoption of sustainable value management for construction projects. For methods where contractors are involved for tendering after the completion of design, it may be easier for the client and his or her representatives to organise the study and incorporate recommendations before tendering. However, except in the case where there is a clause in the contract agreement that sustainable value management should be carried out or that the contracting organisation is also interested in the study, it may be difficult to implement the study for projects in which contractors have been involved from the onset using such procurement methods as turnkey and BOT.

Project Tendering Method

There are various versions of tendering methods available for construction projects. They are classified as competitive or non-competitive and are also referred to as open or closed respectively. A closed tendering method may be preferred if there is a contractor that has been previously involved in a sustainable project. Otherwise, it becomes an issue if an open tendering method is to be adopted—owing to such factors as public interest, and transparency—as some of the contractors who will tender for the job may not have adequate knowledge or experience of sustainable goals. However, the knowledge and experience of sustainable construction can be included as one of the prequalification requirements to tender for the job.

Project Contractual Arrangement and Conditions

It has been noted that old (seminal) contract conditions such as Joint Contract Tribunal (JCT) are still in used in some countries as the basis for the execution of construction projects. This will pose a barrier to the adoption and proper implementation of sustainable value management as some of the necessary clauses to enforce these are not present in some of these conditions of contract.

Other Hindrances

Apart from the highlighted and discussed barriers, other factors that may hinder the adoption and application of sustainable value management for construction project include the following:

- Lack of or inadequate training and education on value management basics, approach and techniques;
- Unstable economy;
- Lack of political will and power;
- Corruption and greediness of consultants and contractors;
- Lack of implementation of recommendations emanating from previous studies;
- Inconsistency in value management basics, approach and techniques;
- Inconsistency in value management terminology and methodology;
- Inability to provide extra funds for value management study;
- Construction industry culture which is more concerned about defeating competitors than concentrating on the demand of clients and consumers; and
- High initial cost of the study.

Summary

After a careful discussion of various limitations and hindrances to the successful initiation and implementation of sustainable value management in this chapter, the next chapter considers different approaches and means of reducing or eliminating these barriers.

References

Abidin, N. Z., & Pasquire, C. L. (2005). Delivering sustainability through value management: Concept and performance overview. *Engineering, Construction and Architectural Management, 2*(2), 168–180.

AlSanad, S. (2015). Awareness, drivers, actions, and barriers of sustainable construction in Kuwait. *Procedia Engineering, 118*(2015), 969–983.

Al-Yami, A. M., & Price, A. D. (2005). Exploring conceptual linkages between value engineering and sustainable construction. In F. Khosrowshahi (Ed.), *21st Annual ARCOM Conference* (pp. 375–384). Association of Researchers in Construction Management: University of London.

Al-Yousefi, A. S. (2016). The synergy between value engineering & sustainability. *Presentation at the 8th World Congress of the Council on Tall Buildings and Urban Habitat (CTBUH)*, Dubai. Retrieved November 14, 2016, from http://www.energyandwateroman.com.

Ametepey, O., Aigbavboa, C., & Ansah, K. (2015). Barriers to successful implementation of sustainable construction in the Ghanaian construction industry. *Procedia Manufacturing, 3* (2015), 1682–1689.

Bowen, P., Cattell, K., Edwards, P., & Jay, I. (2010a). Value management practice by South African quantity surveyors. *Facilities, 28*(1/2), 46–63.

Bowen, P., Jay, I., Cattell, K., & Edwards, P. (2010b). Value management awareness and practice by South African architects: An empirical study. *Construction Innovation, 10*(2), 203–222.

Daddow, T., & Skitmore, M. (2005). Value management in practice: An interview survey. *Australian Journal of Construction Economics and Building, 4*(2), 11–18.

Jay, C. I., & Bowen, P. I. (2015). Value management and innovation: A historical perspective and review of the evidence. *Journal of Engineering, Design and Technology, 13*(1), 123–143.

Karunasena, G., Rathnayake, R. M., & Senarathne, D. (2016). Integrating sustainability concepts and value planning for sustainable construction. *Built Environment Project and Asset Management, 6*(2), 125–138.

Leung, M.-Y., & Yu, J. (2014). Value methodology in public engagement for construction development projects. *Built Environment Project and Asset Management, 4*(1), 55–70.

Liu, G., & Shen, Q. (2005). Value management in China: Current state and future prospect. *Management Decision, 43*(4), 603–610.

Oke, A. E. & Ogunsemi, D. R. (2011). Value management in the Nigerian construction industry: Militating factors and the perceived benefits. *Proceedings of the 2nd International Conference on Advances in Engineering and Technlology* (pp. 353–359). Faculty of Technology, Makerere University, Uganda.

Oke, A. E., Aghimien, D. O., & Olatunji, S. O. (2015). Implementation of value management as an economic sustainability tool for building construction in Nigeria. *International Journal of Managing Value and Supply Chains, 6*(4), 55–64.

Perera, S., Hayles, C. S., & Kerlin, S. (2011). An analysis of value management in practice: The case of Northern Ireland's construction industry. *Journal of Financial Management of Property and Construction, 16*(2), 94–110.

Roufechael, K. M., Abu Bakar, A. H., & Tabassi, A. A. (2015). Value management and client attitude in developing sustainable construction. *Advances in Environmental Biology, 9*(5), 4–6.

Shen, Q., & Liu, G. (2004). Applications of value management in the construction industry in China. *Engineering, Construction and Architectural Management, 11*(1), 9–19.

Yekini, A. A., Bello, S. K. & Olaiya, K. A. (2015). Application of value engineering techniques in sustainable product and service design. *Science and Engineering Perspectives, 10*(Sept), 120–130.

Chapter 11
Drivers of Sustainable Value Management

Abstract There are many benefits of applying sustainable value management for construction projects but several barriers have also been identified. This chapter explains various means of improving the willingness to apply and employ a sustainable value management study for construction projects. Drivers highlighted and discussed include training and education, creating the necessary awareness, the involvement of stakeholders, and the formulation of the necessary and appropriate guidelines and regulations, among others. The drivers discussed in this chapter will help in promoting and adopting the study when properly considered and implemented accordingly.

Keywords Construction project · Research and development · Training · Stakeholder management · Sustainable construction · Sustainable value management · Value management

Introduction

This chapter is a follow-up to Chap. 10 where various barriers to the adoption and implementation of sustainable value management were highlighted and discussed. This chapter considers various means of improving the willingness to apply and employ the study for construction projects.

Various actions can be adopted as drivers for sustainable value management (The College of Estate Management 1995; Male 2002; Abidin and Pasquire 2005; Liu and Shen 2005; Bowen et al. 2010; Perera et al. 2011; Shen and Liu 2004; Leung and Yu 2014; Aghimien and Oke 2015; AlSanad 2015; Jay and Bowen 2015; Noor et al. 2015; Roufechael et al. 2015; Chang et al. 2016). These include training and education, creating the necessary awareness, the involvement of stakeholders, and the formulation of the necessary and appropriate guidelines and regulations.

A.E. Oke and C.O. Aigbavboa, *Sustainable Value Management for Construction Projects*, DOI 10.1007/978-3-319-54151-8_11

Stakeholders Readiness and Participation

The willingness and readiness of various stakeholders in the construction industry to adopt value management that incorporates sustainable goals is fundamental to the adoption and usage of the study. The external and potential internal stakeholders as identified in Chap. 9 are required to play their role in this regard. There is a need for knowledgeable professionals to disseminate and share knowledge about the study and its benefits to construction projects with others.

Adequate Policy, Regulations and Guidelines

This is concerned with the need for laws, regulations, rules, and policies to support the practice of sustainable value management by governments, regulatory bodies, professional bodies, and environmental agencies, who are concerned with the monitoring and controlling of construction projects activities and associated professional services. Moreover, there is a need for guidelines on the implementation of the practice to be made available to stakeholders in the construction industry, including clients and professional bodies.

Awareness and Knowledge of the Practice

There is a need to create awareness of sustainable construction and value management among various concerned stakeholders, including clients and construction professionals in both the public and private sectors of the economy. Research and development are also vital in understanding the people in a particular area, their perceptions of the practice and how the concepts can be promoted.

Training and Education

One of the important drivers of the promotion of a new concept is through training and education of concerned stakeholders on the basics, techniques, and benefits of the concept. The issue of sustainability is still new in some developing countries and value management has not been fully adopted for construction projects in most of their construction projects. It is therefore necessary to ensure that the curricula of higher institutions offering construction-related courses are updated to include these concepts and other emerging ones. Moreover, sustainability and value management should form part of the competencies, duties, and required skills for members of the professional bodies. Prospective members of these bodies will then be tested and examined on these. Various workshops, seminars, conferences, and the like that are organised for the professional development of existing members should also include these practices as well as guidelines on how they can be practically adopted for construction projects.

Societal Awareness

Community participation is fundamental to sustainable development and there is a need for people to be aware of the basics of the practice and their benefits to their quality of life, their environment, and the fact that it is not just about them but future generations as well. This becomes more important for public projects as well as those with a major impact on the economic and social life of the people. The community leaders and other notable members of society can also be included in the sustainable value management study for further engagement in, participation in, and consensus on major project issues.

Modern Approaches to Value Management Workshops

Some new and upcoming teamwork approaches can be adopted for sustainable value management practice and study. These include the following:

- Electronic value management;
- Virtual team approach;
- Delphi techniques; and
- Shortened workshop.

Other Drivers

Apart from the listed and discussed drivers, the following are also essential to the use and adoption of sustainable value management for construction projects:

- Establishing group support system on value management;
- Offering economic incentives for projects that adopt sustainable goals; and
- Developing comprehensive data bases on construction sustainability and value management.

Summary

There are many benefits to applying sustainable value management for construction projects but several barriers have also been identified. The drivers discussed in this chapter will help in promoting and adopting the study when properly considered and implemented accordingly.

References

Abidin, N. Z., & Pasquire, C. L. (2005). Delivering sustainability through value management: Concept and performance overview. *Engineering, Construction and Architectural Management, 2*(2), 168–180.

Aghimien, D. O., & Oke, A. E. (2015). Application of value management to selected construction projects in Nigeria. *Developing Country Studies, 5*(17), 8–14.

AlSanad, S. (2015). Awareness, drivers, actions, and barriers of sustainable construction in Kuwait. *Procedia Engineering, 118*(2015), 969–983.

Bowen, P., Jay, I., Cattell, K., & Edwards, P. (2010). Value management awareness and practice by South African architects: An empirical study. *Construction Innovation, 10*(2), 203–222.

Chang, R., Soebarto, V., Zhao, Z., & Zillante, G. (2016). Facilitating the transition to sustainable construction: China's policies. *Journal of Cleaner Production, 131*(2016), 534–544.

Jay, C. I., & Bowen, P. I. (2015). Value management and innovation: A historical perspective and review of the evidence. *Journal of Engineering, Design and Technology, 13*(1), 123–143.

Leung, M.-Y., & Yu, J. (2014). Value methodology in public engagement for construction development projects. *Built Environment Project and Asset Management, 4*(1), 55–70.

Liu, G., & Shen, Q. (2005). Value management in China: current state and future prospect. *Management Decision, 43*(4), 603–610.

Male, S. (2002). A re-appraisal of value methodologies in construction. *Construction Management and Economics, 11*(2002), 57–75.

Noor, N. F., Kamruzzaman, S. N., & Ghaffar, N. (2015). Sustainability concern in value management: A study on government's building project. *International Journal of Current research and Academic Review, 2*(2015), 72–83.

Perera, S., Hayles, C. S., & Kerlin, S. (2011). An analysis of value management in practice: The case of Northern Ireland's construction industry. *Journal of Financial Management of Property and Construction, 16*(2), 94–110.

Roufechael, K. M., Abu Bakar, A. H., & Tabassi, A. A. (2015). Value management and client attitude in developing sustainable construction. *Advances in Environmental Biology, 9*(5), 4–6.

Shen, G. Q., & Yu, A. T. (2012). Value management: Recent developments and way forward. *Construction Innovation, 12*(3), 264–271.

Shen, Q., & Liu, G. (2004). Applications of value management in the construction industry in China. *Engineering, Construction and Architectural Management, 11*(1), 9–19.

The College of Estate Management. (1995). *Value engineering*. Retrieved May 12, 2016, from http://www.cem.ac.uk/postalcourses.

Chapter 12
Benefits of Sustainable Value Management

Abstract The combination of the sustainability concept and value management for construction projects will benefit not only the client and project but will also help in generating ideas that will improve the environment and quality of lives of people now and in the future. The benefits include those of sustainable goals and objectives of value management practice. This chapter discusses various direct and indirect benefits resulting from the adoption and application of sustainable value management in construction. There are numerous benefits of sustainable value management but they can only be harnessed if adopted, if the right mix of people is engaged, and if recommendations from the study are implemented accordingly. These therefore become the joint roles of clients, consultants, and other construction project stakeholders in harnessing the benefits of sustainable value management.

Keywords Construction project · Innovativeness · Sustainable construction · Sustainable value management · Unnecessary cost · Value for money · Value management

Introduction

The combination of the sustainability concept and value management for construction projects will benefit not only the client and project but will also help in generating ideas that will improve the environment and quality of lives of people now and in the future. The benefits include those of sustainable goals and objectives of value management practice. This chapter discusses various direct and indirect benefits resulting from the adoption and application of sustainable value management in construction.

A.E. Oke and C.O. Aigbavboa, *Sustainable Value Management for Construction Projects*, DOI 10.1007/978-3-319-54151-8_12

General Project Benefits

This section discusses the general benefits of sustainable value management to construction projects (Bourdeau 1999; Abidin and Pasquire 2005; Al-Yami and Price 2005; Daddow and Skitmore 2005; Short et al. 2008; Jay and Bowen 2015; Oke et al. 2015; Roufechael et al. 2015; Ruparathna and Hewage 2015; Yekini et al. 2015; Choi et al. 2016; Karunasena et al. 2016; Alwan et al. 2017). These include the following:

- Better and clearer focus of project objectives prior to construction stage;
- Benefit of meeting required objectives and needs for the project while prioritizing necessary ones;
- Beneficial to overall project management of construction projects where the principle is adopted and implemented;
- Benefit of challenging personal views of project team members to arrive at a common project objectives;
- Benefit of effective and efficient design for project success;
- Benefit of waste reduction;
- Benefits of alternative sustainable construction methods;
- Aid discovery of and adjustments to new materials and techniques for sustainable construction;
- Early discovery and discussion of risks and constraints to project success before the actual construction phase;
- Aid difficult project decision-making through the involvement of appropriate stakeholders;
- Benefit of identifying and removing of unnecessary costs, thereby saving money for clients and financiers;
- The benefit of whole-life cost at the initial stage of project helps in better selection of alternatives and investment appraisal;
- Benefit of ascertaining profitability of projects at an early stage for appropriate decision-making by concerned stakeholders;
- Benefits of supporting information and documents required for sourcing for sponsors and financiers for construction project;
- Benefits of incorporating necessary stakeholders for informed and widely acceptable decisions for construction projects;
- Benefit of considering various and numerous alternatives and ideas before making decisions;
- Benefit of achieving maximum efficiency and effectiveness of project objectives and goals;
- Benefit of addressing specifications issues at the early stage of projects to meet require standards and objectives;
- Benefit of planning early which can save cost and time of delivering the project if properly implemented;
- Benefit of improving functional space quality of construction projects;

- Benefits of early identification of problems and issues that may affect planned delivery of projects;
- Benefit of authoritative evaluation of project brief and design for management decisions;
- Benefit of adaptability and flexibility of project goals and objectives;
- Benefit of delivering project in the most cost-effective way; and
- Benefit of involvement of client and other stakeholders at the preconstruction stage to better harness/identify their needs and requirements for a better definition of project scope and objectives.

Innovations

Sustainable value management adopts the principles and objectives of sustainability and value management simultaneously in decision making, with a view to improving existing materials, methods, and techniques for better construction projects. In proposing alternatives, the teamwork approach of individuals with diverse skills, experience, and perceptions will not only lead to numerous ideas but also the introduction of new, pioneering, ground-breaking, and original ones. After careful evaluation, it is expected that the best alternative idea will not only improve designs and the project itself but will also enhance innovative projects that will be a reference for others. Sustainable value management will therefore challenge the status quo and help in developing innovative design solutions.

Elimination of Unnecessary Cost

The concept of unnecessary cost and their constituents was discussed in chapter three. This is the concern of the economic element of sustainable principle and also comprises a major principle in value management which targets achieving a function at the lowest possible cost. There are also unnecessary costs as a result of the adoption of non-renewable and non-recyclable materials, waste, legal issues, and projects only useful for the present but not suitable for the future. The adoption of sustainable value management will aid in reducing and eliminating unnecessary costs in all aspects of construction projects, right from the preconstruction stage, to construction, re-use, and demolition. This will further aid the efficient use of resources, especially raw materials for construction project.

Whole-Life Cycle Consideration

The economic aspect of sustainable developments considers the financial implications of a project or an element which is inclusive of all relevant costs. This is the same for the whole-life costing principle in value management. Moreover, the other aspects of social and environmental sustainability goals are not only considered for the present but also for the future, which relates to the whole-life concept as well. In view of this, sustainable value management, if effectively and efficiently conducted, will help in considering issues regarding the entire lifespan of construction projects. This will affect the design and type of materials but with the emphasis on flexible sustainable goals without compromising other basic requirements.

Early Problem Identification

It has been discussed in chapters two, three and eight that it is better for sustainable value management to be introduced to a construction project at the preconstruction stage, especially before the completion of designs. This is to ensure that recommendations from the study can be planned and incorporated into the project. This arrangement is beneficial in that through team work, brainstorming, data collection, and evaluation of alternatives, any problems with elements, parts, components, or the entire project are discovered early in the project and the means of addressing them are also specified. This is one of the principles of risk management on which value management excels, the advantage is that the necessary measures or risk response techniques can be put in place, either to prevent the occurrence or prepare for it when it eventually happens.

Adoption of New Materials and Technologies

In the quest for sustainable materials that serve the identified project functions in the same or better way at the least possible cost, it is always the case that new materials and technologies are discovered and recommended by the sustainable value management team. This will benefit project sustainability and make it relevant, not only in the present but also in the future.

Value for Money

A major objective of construction projects is the ability of the project to recoup its initial capital and generate a return on investment as soon as possible. Without compromising sustainability, quality, standards, or other objectives, sustainable value management helps in achieving value for money for project stakeholders, including clients, owners, financiers, end-users, customers, and people in the environment of the project as well as those who will be in the future. Investment appraisals of alternatives are also carried out and presented to the client and project team as the basis for the selection of the best alternative proposal.

Project Clients' and Owners' Benefits

The major benefits of sustainable value management should be for the project for which it was adopted, applied, and commissioned. However, there are also some benefits of the study to the client organisation, owner, sponsor, and financier of the project, regardless of who is responsible for the commissioning of the study (Gan et al. 2015; Al-Yousefi 2016). These benefits include the following:

- Improvement of overall construction project performance;
- Ability to arrive with diverse and many practical solutions to identified construction problems at the early stage and within the shortest possible time;
- Improvement of project designs to meet the needs of clients and other basic requirements;
- Identifying, specifying and defining project scope, quality, and expectations early for proper planning;
- Savings on the cost, time, and effort on the original design and plan;
- Identifying risks as well as highlighting ways to respond; and
- Addressing complicated, complex, and knotty construction issues.

Benefits to Sustainable Value Management Team

There are some indirect benefits of a sustainable value management study to the team as a whole and as well as to individual members. To the team, the following are among the benefits:

- The ability to combine sustainable goals and value management objectives will help in the better delivery of goals and objectives for which the study was commissioned;
- It provides a platform and structure for effective team collaboration;
- Teamwork provides a partnering platform with its associated benefits;
- Consensus and mutual understanding among members of the team are enhanced;
- Communication and team spirit are improved, regardless of professional discipline, experience, training or status of members;

- There is an improved focus on project objectives and meeting them through teamwork;
- It allows for a clear definition of the roles and responsibilities of each member of the team;
- It helps in ensuring a joint ownership of ideas for project implementation;
- It aids effectiveness and efficiency owing to the involvement of members with different team roles, experience and specialisations; and
- It creates an awareness of the team and practice among project stakeholders.

Benefits to Members of Sustainable Value Management Team

These include benefits to each of the participants in sustainable value management including the facilitators, construction professionals, non-professionals, and others (Aghimien and Oke 2015; Al-Yousefi 2016). The benefits to members who participated in sustainable value management include the following:

- It will help members who intend to become value management facilitators with the basic experience required for certification;
- It will increase their sustainable construction awareness and knowledge;
- It will help them in incorporating sustainability goals in their subsequent professional services to their clients;
- It will help in correcting their views and perceptions of sustainable goals and value management;
- It will enhance their ability to participate in further studies with enhanced contribution;
- It will enhance the innovative management techniques of members;
- It will increase the decision-making ability of members;
- It helps in fostering relationships among participants for future collaboration;
- It will help in the understanding of views and perceptions of diverse people in developing their people management skills;
- It helps members to be systematic in solving construction and other life issues;
- It will aid the ability to differentiate between alternative ideas in construction as well as in their personal lives for the purpose of selecting the best one; and
- It will enhance team work participation of members for subsequent value management or other team work management engagement.

Other Benefits

Apart from the three identified areas, other benefits of adopting and incorporating sustainable value management for construction projects include the following:

- Encourages the use of local materials;
- Reduces project abandonment;
- Aids the effective and efficient utilization of resources;
- Enhances the competitive edge for the project contractor;
- Enhances the indoor environmental quality of project;
- Improves the quality of life of users, occupants, or project owners; and
- Promotes a safe environment for current and future generations of people among others.

Summary

There are numerous benefits of sustainable value management but they can only be harnessed if adopted, if the right mix of people is engaged, and if the recommendations from the study are implemented accordingly. These therefore become the joint roles of clients, consultants, and other construction project stakeholders to harness the benefits of sustainable value management.

References

Abidin, N. Z., & Pasquire, C. L. (2005). Delivering sustainability through value management: Concept and performance overview. *Engineering, Construction and Architectural Management, 2*(2), 168–180.

Aghimien, D. O., & Oke, A. E. (2015). Application of value management to selected construction projects in Nigeria. *Developing Country Studies, 5*(17), 8–14.

Alwan, Z., Jones, P., & Holgate, P. (2017). Strategic sustainable development in the UK construction industry, through the framework for strategic sustainable development, using building information modelling. *Journal of Cleaner Production, 140*(2017), 349–358.

Al-Yami, A. M. & Price, A. D. (2005). Exploring conceptual linkages between value engineering and sustainable construction. In F. Khosrowshahi (Ed.), *21st Annual ARCOM Conference* (pp. 375–384). University of London: Association of Researchers in Construction Management.

Al-Yousefi, A. S. (2016). *The synergy between value engineering & sustainability*. Presentation at the 8th World Congress of the Council on Tall Buildings and Urban Habitat (CTBUH), Dubai. Retrieved November 14, 2016, from http://www.energyandwateroman.com.

Bourdeau, L. (1999). Sustainable development and the future of construction: A comparison of visions from various countries. *Building Research & Information, 27*(6), 354–366.

Choi, S. W., Oh, B. K., Park, J. S., & Park, H. S. (2016). Sustainable design model to reduce environmental impact of building construction with composite structures. *Journal of Cleaner Production, 137*(2016), 823–832.

Daddow, T., & Skitmore, M. (2005). Value management in practice: An interview survey. *Australian Journal of Construction Economics and Building, 4*(2), 11–18.

Gan, X., Zuo, J., Ye, K., Skitmore, M., & Xiong, B. (2015). Why sustainable construction? Why not? An owner's perspective. *Habitat International, 47*(2014), 61–68.

Jay, C. I., & Bowen, P. I. (2015). Value management and innovation: A historical perspective and review of the evidence. *Journal of Engineering, Design and Technology, 13*(1), 123–143.

Karunasena, G., Rathnayake, R. M., & Senarathne, D. (2016). Integrating sustainability concepts and value planning for sustainable construction. *Built Environment Project and Asset Management, 6*(2), 125–138.

Oke, A. E., Aghimien, D. O., & Olatunji, S. O. (2015). Implementation of value management as an economic sustainability tool for building construction in Nigeria. *International Journal of Managing Value and Supply Chains, 6*(4), 55–64.

Roufechael, K. M., Abu Bakar, A. H., & Tabassi, A. A. (2015). Value management and client attitude in developing sustainable construction. *Advances in Environmental Biology, 9*(5), 4–6.

Ruparathna, R., & Hewage, K. (2015). Sustainable procurement in the Canadian construction industry: Current practices, drivers and opportunities. *Journal of Cleaner Production, 109* (2015), 305–314.

Short, C. A., Barett, P., Dye, A., & Sutrisana, M. (2008). Impacts of value engineering on five capital arts projects. *Construction Management and Economics, 35*(3), 287–315.

Yekini, A. A., Bello, S. K. & Olaiya, K. A. (2015). Application of value engineering techniques in sustainable product and service design. *Science and Engineering Perspectives, 10*(Sept), 120–130.

Index

A.E. Oke and C.O. Aigbavboa, *Sustainable Value Management for Construction Projects*, DOI 10.1007/978-3-319-54151-8

Printed in the United States
By Bookmasters

Printed in the United States
By Bookmasters